Chemical Properties and Analysis of Milk

Chemical Properties and Analysis of Milk

Dr. S.K. Singh

Oxford Book Company
Jaipur ▪ New Delhi

Published By

Oxford Book Company

Regd. Off. "Bony Residency", Gate No. 2
Opp. Tilak Public School, Vishweswariya Nagar,
Gopalpura Road, Jaipur - 302018, India
Ph: 0141 - 2761280, Telefax: 2761381
email - oxfordbookjpr@gmail.com
website - oxfordbookcompany.com

Delhi Office: 102, 1st Floor, Satyam House,
4327/3, Ansari Road, Darya Ganj,
New Delhi - 110002, India
Ph: 011-45652440

Distributed By

KEYA INTERNATIONAL PRESS LLP
E-mail: keyainternationalpressllp@gmail.com

Published in India, 2025

ISBN : 9789350305409

Typeset by Shivangi Computers, Jaipur

Printed and bound in India by Thomson Press Ltd.,
New Delhi

Contents

Preface

The role of milk in nature is to nourish and provide immunological protection for the mammalian young. Milk has been a food source for humans since prehistoric times; from human, goat, buffalo, sheep, yak, to the focus of this section-domesticated cow milk. Milk and honey are the only articles of diet whose sole function in nature is food. It is not surprising, therefore, that the nutritional value of milk is high. Milk is also a very complex food with over 100,000 different molecular species found. There are many factors that can affect milk composition such as breed variations.

Milk from different animals varies somewhat in the proportion of the different constituents. All milk contains a high proportion of water, cow's milk averaging about 87 per cent. The milk from different breeds of cattle varies considerably in composition. The milk of individual cows of the same breed also varies in the proportion of the different constituents. This may be partly due to environment, partly to inheritance, and partly to individuality. Milk may vary particularly in fat content from one milking to the next. The percentage of fat increases during the milking period; that is, the first milk obtained is not so rich in fat as the last portions of milk.

The fat varies for different breeds from about 3.5 per cent for Holstein to about 5 per cent for Guernsey and Jersey. The protein varies from about 3.3 per cent for Holstein to 4.0 per cent for Guernsey and Jersey. It parallels the fat content somewhat, being highest in the breeds having a high fat content and lowest in those having a low fat content. The lactose does not vary so much with the different breeds as the fat and protein. The lactose content is from about 4.65 to 5 per cent. The ash varies from about 0.68 to 0.75 for the different breeds.

The constituents of milk that are most important in food preparation are enzymes, vitamins, pigments, salts, sugar, fat, and proteins. Enzymes. The enzymes of cow's milk are reported as follows by Rogers; proteinases, lactase, diastase, lipase, salolase, catalase, peroxidase, and aldehydrase. Rogers states that the proteolytic enzyme, galactase, brings about slow decomposition of milk proteins into peptones, amino acids, and ammonia. Vitamins. All the vitamins recognized at the present time are contained in milk. Some are present in comparatively large and others in smaller amounts.

The appearance of milk is white. This is due to light rays reflected by the colloidally dispersed constituents of the milk, the calcium caseinate, and calcium phosphate. Milk contains two classes of yellow or orange pigments. The water-soluble pigment, which imparts a yellow colour with a green fluorescence to the whey of milk, was formerly called lactochrome. A name recently suggested for this pigment is lactoflavin. It is regarded as one flavin of a specific group, collectively to be called lyochromes. It is possible that lacto-flavin is composed of more than one pigment. Rogers says "lactoflavin forms compounds with saccharides, proteins, and purines. These compounds possibly either occur naturally in milk or readily form during isolations, thus accounting for the several lactoflavins isolated from milk." It is probable that the pigment lactoflavin is one of the five fractions of vitamin G (B2). Milk is relatively rich in this vitamin.

The book is a unique resource for the students and teachers of food hygiene food safety and quality control in a wide range.

— Dr. S.K. Singh

1

Dairy Microbiology: Bacteria, Yeasts and Molds

Food spoilage is an enormous economic problem worldwide. Through microbial activity alone, approximately one-fourth of the world's food supply is lost. Milk is a highly nutritious food that serves as an excellent growth medium for a wide range of microorganisms. The microbiological quality of milk and dairy products is influenced by the initial flora of raw milk, the processing conditions, and post-heat treatment contamination. Undesirable microbes that can cause spoilage of dairy products include Gram-negative psychrotrophs, coliforms, lactic acid bacteria, yeasts, and molds.

In addition, various bacteria of public health concern such as Salmonella spp., Listeria monocytogenes, Campylobacter jejuni, Yersinia enterocolitica, pathogenic strains of Escherichia coli and enterotoxigenic strains of Staphylococcus aureus may also be found in milk and dairy products. For this reason, increased emphasis should be placed on the microbiological examination of milk and dairy foods.

Microbiological analyses are critical for the assessment of quality and safety, conformation with standards and specifications, and regulatory compliance.

In Hungary, a very comprehensive set of microbiological reference criteria covering virtually all foods and including limits for an extensive range of indicators and pathogens was issued by the Ministry of Health in 1998. This official document entitled "Decree No. 4/1998 (XI. 11.) EüM on the acceptable levels of microbiological contamination in foods" is based on ICMSF and FAO principles in that it includes sampling plans and identifies the criteria for acceptance or rejection of a lot. Altogether 26 food groups have mandatory criteria and the number or type of microorganisms for which there are limits for any particular food varies from one to nine.

Divided into nine subgroups, milk and milk products constitute one of the 26 groups formed. As of 2003, in terms of dairy foods, this legal document was replaced by another one titled "Decree No. 1/2003 (I. 8.) FVM-ESzCsM on the food hygienic conditions for production and placing on the market of raw milk, heat-treated milk and milk-based products", which is identical in content with the EU Milk Hygiene Directive 92/46/EEC.

Production and use of Microbial Enzymes for Dairy Processing

Indians are known to be lovers of milk and its products. As a sequel to white revolution, India has surged ahead to become the largest milk producer in the world, the production figure touched to 74 million tons for the year 1998. With growing urbanization, demand for processed dairy foods has increased considerably, in particular demand for different cheese varieties, and low-lactose milk due to increasing intolerance of human beings to lactose in milk and other milk products. For improving the quality of milk and milk products, a number of different enzymes from microbial as well as from nonmicrobial sources have potential applications in dairy processing.

The use of rennet in cheese manufacture was among the earliest applications of exogenous enzymes in food processing, dating back to approximately 6000 B C. In 1994, the total production of cheese was 8000 metric tons against a total demand of 9000 metric tons. The projected demand by 2000 A D is around 30,000 metric tons. The use of rennet, as an exogenous enzyme, in cheese manufacture is perhaps the largest single application of enzymes in food processing.

In recent years, proteinases have found additional applications in dairy technology, for example in acceleration of cheese ripening, modification of functional properties and preparation of dietic products. Principal among some enzymes that have important and growing applications are lipases and b-galactosidases. Enzymes with limited applications include glucose oxidase, superoxide dismutase, sulphydryl oxidase, etc. The increasing use of enzymes to produce specific products with characteristic attributes can be emphasized by the world-wide sale of industrial enzymes approximating to US $ 1.6 billion, which is expected to reach US $ 3.0 billion by the year 2008. Almost 45% of this is shared by the food industry and the remaining is shared by detergent (34.4%), textile (11%), leather (2.8%), pulp and paper (1.2%) industries and other industries (5.6%) excluding enzymes for use in diagnostics and therapeutics.

India being the highest producer of milk in the world, and consequently the surplus availability of milk in our country has triggered the food and dairy industry to convert the liquid milk into value-added products using biochemical and enzymatic processes.

Microbial Rennets in Dairy Applications

Animal rennet (bovine chymosin) is conventionally used as a milk-clotting agent in dairy industry for the manufacture of quality cheeses with good flavour and texture. Owing to an increase in demand for cheese production world wide – i.e. 4% per annum over the past 20 years, approximating 13.533 million tons – coupled with reduced supply of calf rennet, has therefore led to a search for rennet substitutes, such as microbial rennets. At present, microbial rennet is used for one-third of all the cheese produced world wide.

Rennin acts on the milk protein in two stages, by enzymatic and by nonenzymatic action, resulting in coagulation of milk. In the enzymatic phase, the resultant milk becomes a gel due to the influence of calcium ions and the temperature used in the process. Many microorganisms are known to produce rennet-like proteinases which can substitute the calf rennet.

Microorganisms like Rhizomucor pusillus, R. miehei, Endothia parasitica, Aspergillus oryzae, and Irpex lactis are used extensively for rennet production in cheese manufacture. Extensive research that has been carried out so far on rennet substitutes has been reviewed by several authors. Different strains of species of Mucor are often used for the production of microbial rennets. Whereas best yields of the milk-clotting protease from Rhizomucor pusillus are obtained from semisolid cultures containing 50% wheat bran, R. miehei and Endothia parasitica are well suited for submerged cultivation. Using the former, good yields of milk-clotting protease may be obtained in a medium containing 4% potato starch, 3% soybean meal, and 10% barley. During growth, lipase is secreted together with the protease. Therefore, the lipase activity has to be destroyed by reducing the pH, before the preparation can be used as cheese rennet.

Microbial rennets from various microorganisms (marketed under the trade names such as Rennilase, Fromase, Marzyme, Hanilase, etc.) being marketed since the 1970s have proved satisfactory for the production of different kinds of cheese. The molecular and enzymatic properties of chymosins have been studied extensively. Although the proteolytic specificities of the three commonly used fungal rennets are considerably different from those of calf chymosin, these rennets have been used to produce acceptable cheeses.

Recently Novo Nordisk has succeeded in expressing just one proteolytic enzyme from the fungus R. miehei in the well-known organism A. oryzae. This host organism is able to produce the single protease that cleaves the ·-casein into a glycomacropeptide and para casein by hydrolysing only at the phe105–met106 peptide bond between phenyl alanine and methionine.

This monocomponent enzyme product has the trade name Novoren. One major drawback of microbial rennet use in cheese manufacture, is the development of off flavour and bitter taste in the nonripened as well as in the ripened cheeses. The rennets from microbial sources are more proteolytic in nature in comparison to rennet from animal sources, resulting in production of some bitter peptides during the process of cheese ripening. Hence, attempts have been made to clone the gene for calf chymosin, and to express it in selected bacteria, yeasts, and molds.

Recombinant Rennets for Cheese Manufacture

Due to shortage of calf stomachs and the economic value of cheese rennet, gene for calf chymosin was one of the first genes for mammalian enzymes that was cloned and expressed in microorganisms. Many different laboratories have cloned the gene for calf prochymosin in Escherichia coli, and analysed the structure of the gene as well as the properties of the recombinant chymosin. The expressed proenzyme in E. coli, is present mainly as insoluble inclusion bodies, comprised of reduced prochymosin as well as molecules that are interlinked by disulphide bridges. After disintegration of the cells, inclusion bodies are harvested by centrifugation.

The individual laboratories have reported some differences in the procedure for renaturation of prochymosin from the inclusion bodies, but all have followed the same general scheme. The enzymatic properties of recombinant E. coli chymosin are indistinguishable from those of native calf chymosin. The enzymes were identical when observed by immunodiffusion in gels, but a slight difference was observed by enzyme linked immunosorbent assay (ELISA).

The gene for prochymosin has also been cloned in Saccharomyces cerevisiae; the levels of expression have been reported to be 0.5 to 2.0% of total yeast protein. In yeast, about 20% of the prochymosin can be released in soluble form which can be activated directly; the remaining 80% is still associated with the cell debris.

The cloning of chymosin was carried out without the prosequence. Though the level of chymosin mRNA was similar to that of prochymosin mRNA, no milk-clotting activity was observed in clones containing the chymosin gene only. The results suggest that the prosequence is essential for correct folding of the polypeptide chain. The zymogen for the aspartic proteases from R. pusillus, also called mucor rennin, has likewise been cloned and expressed in yeast. Studies on its conversion to active form showed that secretion of R. pusillus protease from recombinant yeast was dependent on glycosylation of the enzyme.

Compared to yeast, the filamentous fungi generally secrete larger quantities of proteins into the culture medium. Furthermore, filamentous fungi secrete the heterologous proteins with correct folding of the polypeptide chain, and process correct pairing of sulphydryl groups. The gene for R. miehei protease has been expressed in A. oryzae. Prochymosin has been expressed in Kluyveromyces lactis, A. nidulans, A. niger, and Trichoderma reesei. In most of the cases, the reported yields of the model systems were about 10–40 mg of enzyme per litre of culture medium.

However, 3.3 g of enzyme per liter has also been achieved. Yeast, Kluyveromyces lactis, has recently been used as an efficient host for the secretion of recombinant chymosin, which has led to the development of large-scale production process for chymosin. If produced on an industrial scale, the yields will perhaps be of the latter magnitude. Most of the companies produce recombinant rennet of cattle calf origin in different microbial hosts, however in India, the major source of milk is buffalo, showing a different composition from that of cow. The rennet from buffalo source has inherent compatibility for clotting buffalo milk. Therefore, National Dairy Research Institute, Karnal, has taken a lead in cloning gene of buffalo chymosin. The buffalo chymosin cDNA has been cloned in E. coli and partial N-terminal sequencing of the purified buffalo chymosin indicates that it is highly homologous to the cattle chymosin.

Several cheese making experiments have been carried out with recombinant chymosin and the general aspects of recombinant chymosin have been dealt with in a report. Since most of the rennet (> 90 %) added to cheese milk is lost in the whey, immobilization would considerably extend its catalytic life. Several rennets have been immobilized, but their efficiency as milk coagulants has been questioned. So, there is a fairly general support for the view that immobilized enzymes cannot coagulate milk properly owing to inaccessability of the Phe–Met peptide bond of k-casein, and that the apparent coagulating activity of immobilized rennets is due to leaching of the enzyme from the support.

Different types of conventional cheeses have been successfully made by using recombinant rennet on an experimental or pilot scale. No major differences have been detected between cheeses made with recombinant chymosins or natural enzymes, regarding cheese yield, texture, smell, taste and ripening. The recombinant chymosins are identical with calf rennet according to the report on biochemical and genetic evidences.

Lactase in Dairy Industry

Lactose, the sugar found in milk and whey, and its corresponding hydrolase, lactase or b-galactosidase, have been extensively researched during the past decade. This is because of the enzyme immobilization technique which has given new and interesting possibilities for the utilization of this sugar. Because of intestinal enzyme insufficiency, some individuals, and even a population, show lactose intolerance and difficulty in consuming milk and dairy products.

Hence, low-lactose or lactose-free food aid programme is essential for lactose-intolerant people to prevent severe tissue dehydration, diarrhoea, and, at times, even death. Another advantage of lactase-treated milk is the increased sweetness of the resultant milk, thereby avoiding the requirement for addition of sugars in the manufacture of flavoured milk drinks. Manufacturers of ice cream, yoghurt and frozen desserts use lactase to improve scoop and creaminess, sweetness, and digestibility, and to reduce sandiness due to crystallization of lactose in concentrated preparations. Cheese manufactured from hydrolysed milk ripens more quickly than the cheese manufactured from normal milk.

Technologically, lactose crystallizes easily which sets limits to certain processes in the dairy industry, and the use of lactase to overcome this problem has not reached its fullest potential because of the associated high costs. Moreover, the main problem associated with discharging large quantities of cheese whey is that it pollutes the environment. But, the discharged whey could be exploited as an alternate cheap source of lactose for the production of lactic acid by fermentation. The whey permeate, which is a by-product in the manufacture of whey protein concentrates, by ultrafiltration could be fermented efficiently by Lactobacillus bulgaricus.

Lactose can be obtained from various sources like plants, animal organs, bacteria, yeasts (intracellular enzyme), or molds. Some of these sources are used for commercial enzyme preparations. Lactase preparations from A. niger, A. oryzae, and Kluyveromyces lactis are considered safe because these sources already have a history of safe use and have been subjected to numerous safety tests. The most investigated E. coli lactase is not used in food processing because of its cost and toxicity problems.

Properties of Lactase

The properties of the enzyme depend on its source. Temperature and pH optima differ from source to source and with the type of particular commercial preparation. Immobilization of the enzymes, method of immobilization, and type of carrier can also influence these optima values. In general, fungal lactase have pH optima in the acidic range 2.5–4.5, and yeast and bacterial lactases in the neutral region 6–7 and 6.5–7.5, respectively.

The variation in pH optima of lactases makes them suitable for specific applications, for example fungal lactases are used for acid whey hydrolysis, while yeast and bacterial lactases are suitable for milk (pH 6.6) and sweet whey (pH 6.1) hydrolysis. Product inhibition, e.g. inhibition by galactose, is another property which also depends on the source of lactase. The enzyme from A. niger is more strongly inhibited by galactose than that from A. oryzae.

This inhibition can be overcome by hydrolysing lactose at low concentrations by using immobilized enzyme systems or by recovering the enzyme using ultrafiltration after batch hydrolysis. Lactase from Bacillus species are superior with respect to thermostability, pH operation range, product inhibition, and sensitivity against high-substrate concentration. Thermostable enzymes, able to retain their activity at 60°C or above for prolonged periods, have two distinct advantages viz. they give higher conversion rate or shorter residence time for a given conversion rate, and the process is less prone to microbial contamination due to higher operating temperature. Bacillus species have a pH optima of 6.8 and a temperature optima of 65°C. Its high activity for skim milk, and less inhibition by galactose has made it suitable for use as a production organism for lactase.

The enzymatic hydrolysis of lactose can be achieved either by free enzymes, usually in batch fermentation process, or by immobilized enzymes or even by immobilized whole cells producing intracellular enzyme.

Although numerous hydrolysis systems have been investigated, only few of them have been scaled up with success and even fewer have been applied at an industrial or semi-industrial level. Several acid hydrolysis systems have been developed to industry-scale level. Large-scale systems which use free enzyme process have been developed for processing of UHT-milk and processing of whey, using K. lactis lactase (Maxilact, Lactozyme). Several commercial immobilized systems have been developed for commercial exploitation. Snamprogretti process of industrial-scale milk processing technology in Italy is one such working systems.

They make use of fibre-entrapped yeast lactase in a batch process, and the milk used is previously sterilized by UHT. For pilot plants, there are three other processes designed and developed to handle milk; (i) by Gist-Brocades, Rohm GmbH (Germany), and (ii) by Sumitomo, Japan. These are continuous processes with short residence times. Processing of whey UF-permeate is accomplished by the system developed by Corning Glass, Connecticut, Lehigh, Valio and Amerace corp. The process by Corning Glass is being applied at commercial scale in the bakers yeast production using hydrolysed whey.

Microbial Enzymes in Accelerated Cheese Ripening

Cheese ripening is a complex process mediated by biochemical and biophysical changes during which a bland curd is developed into a mature cheese with characteristic flavour, texture, and aroma.

The desirable attributes are produced by the partial and gradual breakdown of carbohydrates, lipids, and proteins during ripening, mediated by several agents, viz. (i) residual coagulants, (ii) starter bacteria and their enzymes, (iii) nonstarter bacteria and their enzymes, (iv) indegenous milk enzymes, especially proteinases, and (v) secondary inocula with their enzymes.

Proteolysis occurs in all the cheese varieties and is a prerequisite for characteristic flavour development that can be regulated by proper use of the above agents. Cheese ripening is essentially an enzymatic process which can be accelerated by augmenting activity of the key enzymes. This has the advantage of initiating more specific action for flavour development compared to use of elevated temperatures that can result in accelerating undesirable nonspecific reactions, and consequently off flavour development.

Enzymes may be added to develop specific flavours in cheeses, for example lipase addition for the development of Parmesan or Blue-type cheese flavours. Attempts to accelerate the multiple secondary flavour-forming-reactions, e.g. Strecker degradation, have been scarce. The pathways leading to the formation of flavour compounds are largely unknown, and therefore the use of exogenous enzymes to accelerate ripening is mostly an empirical process.

Studies on the hydrolysis of whole casein and isolated casein components to observe the kinetics and specificity of aspartic proteases in rennin, pepsin, and four microbial rennet substitutes indicated the considerable differences in the reaction velocity and the extent of hydrolysis on the rennet curd yield. The rennet enzymes were active even in the later phases of cheese-ripening.

Proteinases and Peptidases

Proteolysis is characteristic of most cheese varieties and is indispensable for good flavour and textural development. Proteinases used in cheese processing include (i) plasmin, (ii) rennet, and (iii) proteinases (cell wall and/or intracellular) of the starter and nonstarter bacteria. Approximately 6% of the rennet added to cheese milk remains in the curd after manufacture and contributes significantly to proteolysis during ripening.

Combinations of individual neutral proteinases and microbial peptidases intensified cheese flavour, and when used in combinations with microbial rennets reduced the intensity of bitterness caused by the latter. Acid proteases in isolation cause intense bitterness. Various animal or microbial lipases gave pronounced cheese flavour, low bitterness and strong rancidity, while lipases in combination with proteinases and/or peptidases give good cheese flavour with low levels of bitterness. In a more balanced approach to the acceleration of cheese ripening using mixtures of proteinases and peptidases, attenuated starter cells or cell-free extracts (CFE) are being favoured.

Proteolytic Enzymes of Lactic Acid Bacteria in Fermented Milk Products

The proteolytic system of lactic acid bacteria is essential for their growth in milk, and contributes significantly to flavour development in fermented milk products. The proteolytic system is composed of proteinases which initially cleaves the milk protein to peptides; peptidases which cleave the peptides to small peptides and amino acids; and transport system responsible for cellular uptake of small peptides and amino acids.

Lactic acid bacteria have a complex proteolytic system capable of converting milk casein to the free amino acids and peptides necessary for their growth. These proteinases include extracellular proteinases, endopeptidases, aminopeptidases, tripeptidases, and proline-specific peptidases, which are all serine proteases. Apart from lactic streptococcal proteinases, several other proteinases from nonlactostreptococcal origin have been reported. There are also serine type of proteinases, e.g. proteinases from Lactobacillus acidophillus, L. plantarum, L. delbrueckii sp. bulgaricus, L. lactis, and L. helveticus. Aminopeptidases are important for the development of flavour in fermented milk products, since they are capable of releasing single amino acid residues from oligopeptides formed by extracellular proteinase activity.

Other Dairy Enzymes

Other enzymes used for dairy food application include: (i) proteases to reduce allergic properties of cow milk products for infants, and (ii) lipases for development of lipolytic flavours in speciality cheeses. The functional properties of milk proteins may be improved by limited proteolysis through the enzymatic modification of milk proteins. An acid-soluble casein, free of off flavour and suitable for incorporation into beverages and other acid foods, has been prepared by limited proteolysis. The antigenicity of casein is destroyed by proteolysis, and the hydrolysate is suitable for use in milk-protein-based foods for infants allergic to cow milk.

Lipolysis makes an important contribution to swiss cheese flavours, due mainly to the lipolytic enzymes of the starter cultures. The characteristic peppery flavour of Blue cheese is due to short-chain fatty acids and methyl ketones. Most of the lipolysis in Blue cheese is catalysed by Penicillium roqueforti lipase, with a lesser contribution from indigenous milk lipase.

The NOVO process for production of enzyme modified cheese (EMC) uses medium-aged cheese which is emulsified, homogenized, and pasteurized, after which 'palatase' (a lipase from R. miehei) is added, with or without a proteinase, and the blend is ripened at a high temperature for one to four days.

The mixture is reheated, a paste results which is suitable for inclusion in soups, dips, dressings, or snack foods. EMC technology has been developed to produce a range of characteristic cheese flavours and flavour intensities, for example swiss, blue, cheddar, provolo-nemor or romano, suitable for inclusion at low levels in many products. The claims that exogenous enzymes are effective in accelerating ripening have not led to their wide-spread use, possibly due to their high cost, difficulties in distributing them uniformly in the curd, and the possible danger of over-ripening the cheese.

The other minor enzymes having limited applications in dairy processing include glucose oxidase, catalase, superoxide dismutase, sulphydryl oxidase, lactoperoxidase, and lysozymes. Glucose oxidase and catalase are often used together in selected foods for preservation. Superoxide dismutase is an antioxidant for foods and generates H_2O_2, but is more effective when catalase is present.

Thermally induced generation of volatile sulphydryl groups is thought to be responsible for the cooked off-flavour in ultra high temperature (UHT) processed milk. Use of sulphydryl oxidase under aseptic conditions can eliminate this defect. The natural inhibitory mechanism in raw milk is due to the presence of low levels of lactoperoxidase (LP), which can be activated by the external addition of traces of H_2O_2 and thiocyanate. It has been reported that the potential of LP-system and its activation enhances the keeping quality of milk. Cow milk can be provided with protective factors by the addition of lysozyme, making it suitable as an infant milk. Lysozyme acts as a preservative by reducing bacterial counts in milk without affecting the

L. bifidus *activity. The scope of application of minor enzymes to milk and milk products has been recently reviewed.*

The global market for the production of microbial enzymes for use in dairy-products manufacture is considerably large, but is being dominated only by a limited number of enzyme producers. In India the microbial dairy enzymes requirement has been very limited till now. However, with the advent of technological processes for the manufacture of different varieties of milk products, such as cheeses by the State Dairy Federations, Co-operatives and Private Dairy Product Manufacturers like Amul, Vijaya, Verka, Dynamix, Nestle, Smith Kline Beechem, etc., the markets for the sale of such products in megacities and towns has been slowly growing for the past two to three years.

Presently, many of these microbial enzymes, such as microbial rennets and other enzymes, are being imported. Hence, there is a scope for the production of enzymes such as microbial rennet, lactase, proteinases, and lipases indigenously. In the near future, the requirement for these enzymes is bound to increase by leaps and bounds, basically due to requirement of value-added dairy products in the country.

Common Tests of Milk and Cream

Milk and dairy products constitute an important item of our food. These products are very suitable for microbial growth. It thus becomes necessary to know the chemistry of milk, its spoilage, method of preservation, and different dairy products where microbes play a positive rather than negative role.

Milk is considered as a complete food and it contains proteins, fat, carbohydrates, minerals, vitamins and water.

It is also a good medium for the growth of microorganisms. It is therefore, important to know the types of microorganisms present in milk, their control and use for beneficial purposes.

Milk contains relatively few bacteria when it is secreted from the udder of an healthy animal. However, during milking operations it gets contaminated from the exterior of the upper and the adjacent areas, dairy untensils, milking machines, thc himds of the milkers from the soil and dust.

In this way bacteria, yeasts and molds got into the milk and constitute the normal flora of milk. The number of contaminants added from various sources depends on the care taken to avoid contamination.

The presence of these nonpathogenic organisms in milk is not serious but if these organisms multiply quickly, they can cause spoilage of milk, such as souring or putrefaction and develop undesirable odours. Control of their multiplication in milk is therefore, very essential.

Milk may also contain pathogenic organisms, derived directly from the animal or from the surroundings. Microorganisms that are harmful and found in milk are Streptococcus cremoris, Pseudomonas sp., Mycobacterium spp.

Serratia marcescens, enteric bacteria etc. Normally, milk is pasteurized before use. However, pasteurization does not kill all the bacteria; the survivors (thermodurics), depending on their initial number then multiply and if the initial number is high they cause rapid spoilage. It is imporant, therefore, that the milk be refrigerated at. around O°C soon after pasteurization to prevent the growth of these undersirable microorganisms.

Pasteurization, either at 145°F for 30 minutes or 161°P for 15"30 seconds eliminates most of the pathogenic bacteria particularly Mycobacterium tuberculosis.

Boiling of milk destroys all microorganisms except spore formers. Sometimes, on cooling or under improper refrigeration, spores germinate and cause spoilage of boiled milk.

Detailed Descriptions: Milk Component Analysis

Analyst Performing Milk Component Test with the MilkoScan™ FT 120

This electronic instrument provides us the ability to quickly analyse milk samples for components (i.e. fat, protein, lactose, total solids and solids non-fat). MilkoScan ™ FT 120 employs the Fourier transform Infra-Red Spectrophotometer (FTIR) measuring principle. In basic terms, the instrument provides milk component results through infrared light measurement. The Foss MilkoScan ™ FT 120 can provide test results in about 30 seconds and can be calibrated for raw bovine and caprine milk, homogenized milks, cream, whey and other dairy products.

Milk Component Chemistry

Babcock Method for Fat : This is relatively inexpensive test for fat that can be performed on raw milk and a limited number of dairy products. This procedure uses sulfuric acid to "burn everything in the milk sample" except for the fat. The remaining fat in the sample is then centrifuged and determined graduated markings on specialized glassware. The procedure is limited in that individual samples can only be read to the 0.05% fat level.

Gerber Method for Fat

This is a test is close relative to the Babcock method. It is often referred to as Babcock's European cousin as this test is most common in the European dairy industry. Milk fat is separated from other milk components by adding sulfuric acid. The separation is facilitated by using amyl alcohol and centrifugation. The fat content is read directly via a special calibrated butyrometer. The procedure is limited in that individual samples can only be read to the 0.05% fat level.

Mojonnier (Ether Extract) Method

This is the chemical method of choice for total milk fat because of its inherent analytical capacity. It can provide a very precise measurement of milk fat. Because of its precision, the ether extraction procedure is the preferred chemical reference method for milk fat. Ether and alcohol are used to extract the fat from milk. The ether extract is decanted into a dry weighing dish and the ether is evaporated. The extracted fat is then dried to a constant weight and is expressed as % fat by weight.

Kjeldahl Method for Protein

This chemical method determines the nitrogen content of milk. Acid digestion of the milk sample occurs at high temperature. Nitrogen content is then converted by known formula to provide the milk protein%.

Direct Forced Air Oven Method for Total Solids

Total solids are determined by weighing milk, drying milk and weighing the dried milk residue. The sample is dried in a forced air oven for 4 hours at 100 + 1°C. The total solids content of milk is the weight of dried milk expressed as a percentage of the original milk weight.

Electronic Somatic Cell Count (ESCC) with the Fossomatic ™ Minor

Electronic instrument used in the detection of Somatic cells in milk. The milk sample is mixed with a staining solution that dyes the DNA in somatic cells. The stained sample is then pipetted into the instrument. Light is used to fluoresce the dyed somatic cells. The fluoresced cells are then counted by using a light sensitive chip.

Direct Microscopic Somatic Cell Count

The SCC is determined by viewing stained somatic cells under a microscope under oil immersion. The analysts' estimated counts are determined by using a microscopic factor based on the field of view of the microscope.

Freezing Point Depression (FPD) of Milk with the Cryoscope

This test is used to detect the adulteration of water in milk. Due to the soluble components of milk (i.e. lactose and salts), normal milk freezes slightly below the "freezing point" of pure water. The cryoscope instrument is used to determine the freezing point of milk. Small amounts of added water dilute the soluble components of milk causing the freezing point to move upward toward that of pure water.

Interpretations of cryoscope readings can be awkward. Years ago, cryoscope readings were reported with a unique measurement scale known at the Hortvet scale (°H). However, today most cryoscope results are reported in Centigrade (°C). In some cases, laboratories have inadvertently calibrated their cryoscopes to °H and reported results as °C. When interpreting cryoscope results, first determine which measurement scale (°H or °C), is being used to calibrate the lab's cryoscope. Then apply the reading to the appropriate scale. Normal comingled loads of milk should have cryoscope values between-0.517 and-0.527 °C (-0.540 and-0.550 °H).

However, normal single herd result may vary depending on a variety of conditions including (but not limited to) herd health, season of the year, mineral content of the herd's diet, animal breed and herd stress. Because of these factors, the concerns implicating adulteration of milk with water usually are noted at-0.503 °C (-0.525 °H). If a herd's cryoscope results consistently show concern, an evaluation of the herd's milk should be conducted to determine a "baseline" for that specific herd's normal cryoscope reading. The cryoscope at Regulatory Services' milk lab is calibrated in °C and all test results are reported in °C.

Standard Plate Count Bacteria (SPC) with Petrifilm®

The Petrifilm® plate is an all-in-one plating system method that provides for the Standard Plate Count (SPC) of aerobic bacteria in milk. Pre-made films of agar are inoculated with 1:100 and 1:1000 dilutions of milk. The inoculated Petrifilm® is incubated for 48 hours and the bacteria colonies are counted by the analyst.

Antibiotic Test with Charm SL®

Detection of antibiotics or other growth inhibitors in raw milk is made possible by validated growth inhibitor assays or by rapid receptor based screening processes. Acceptable test methods have been validated and approved by FDA are identified in PMO related documents. The Charm SL® is approved for the determination of antibiotics in raw cow and goat milk. The Charm SL® is a receptor assay utilizing rapid one step assay lateral flow technology. The procedure detects Beta-Lactam drugs (Amoxicillin, Ampicillin, Ceftiofor, Cephapicin, Penicillin G) at or below safe level/tolerance.

Sediment

The sediment test determines the level of particulates present in milk and is an indicator of milk production practices and cleanliness. Sediment is determined by filtering a portion of milk by suction through a fine fibre filter. The filter is then visually examined to determine the milligrams of sediment present per gallon. In years past, the sediment test was quite common throughout the dairy industry. Improvements in farm production practices and in milk quality have diminished the need for routine sediment testing in most parts of the US.

Other Tests

The milk laboratory at Regulatory Services has the capacity to perform the milk analyses identified above. Periodically we may add new procedures. Please feel free to contact us for milk testing inquiries not identified in this document.

Milk Testing and Quality Control

Milk testing and quality control is an essential component of any milk processing industry whether small, medium or large scale. Milk being made up of 87% water is prone to adulteration by unscrupulous middlemen and unfaithful farm workers. Moreover, its high nutritive value makes it an ideal medium for the rapid multiplication of bacteria, particularly under unhygienic production and storage at ambient temperatures. We know that, in order for any processor to make good dairy products, good quality raw materials are essential. A milk processor or handler will only be assured of the quality of raw milk if certain basic quality tests are carried out at various stages of transportation of milk from the producer to the processor and finally to the consumer. There are a number of standard manuals and text books on milk quality control. However these may not be easily available to the emerging small scale to medium scale processors in Kenya.

For these reasons, the Training Programme for the Small Scale Dairy Sector under project GOK/FAO/TCP/KEN/6611, has assembled this guide on Milk Testing and Quality Control so that it may be used for training and by the private small scale dairy processors. The methods selected are simple and basic and will suffice the requirements of most milk quality control laboratories of small scale processing units. For the larger plants with bigger laboratories more tests are to be found in the bibliography at the end of this booklet.

Milk Testing And Quality Control

What is Milk Quality Control?

Milk quality control is the use of approved tests to ensure the application of approved practices, standards and regulations concerning the milk and milk products. The tests are designed to ensure that milk products meet accepted standards for chemical composition and purity as well as levels of different micro-organisms.

Why have Milk Quality Control?

Testing milk and milk products for quality and monitoring that MILK PRODUCTS, PROCESSORS and MARKETING AGENCIES adhere to accepted codes of practices costs money. There must be good reasons why we have to have a quality control system for the dairy industry in Kenya.

The reasons are:

- *To the Milk Producer:* The milk producer expects a fair price in accordance with the quality of milk she/he produces.
- *The Milk Processor:* The milk processor who pays the producer must assure himself/herself that the milk received for processing is of normal composition and is suitable for processing into various dairy products.
- *The Consumer:* The consumer expects to pay a fair price for milk and milk products of acceptable to excellent quality.
- The Public and Government Agencies. These have to ensure that the health and nutritional status of the people is protected from consumption of contaminated and sub-standard foodstuffs and that prices paid are fair to the milk producers, the milk processor and the final consumer.

All the above-is only possible through institution of a workable quality testing and assurance system conforms to national or internationally acceptable standards.

Quality Control in the Milk Marketing Chain in Kenya

i) At the farm. Quality control and assurance must begin at the farm. This is achieved through farmers using approved practices of milk production and handling; and observation of laid down regulations regarding, use of veterinary drugs on lactating animals, regulations against adulterations of milk etc.

ii) At Milk collection Centres. All milk from different farmers or bulked milk from various collecting centres must be checked for wholesomeness, bacteriological, and chemical quality.

iii) At the Dairy Factories. Milk from individual farmers or bulked milk from various collecting centres

iv) Within the Dairy Factory. Once the dairy factor has accepted the farmer milk it has the responsibility of ensuring that the milk is handled hygienically during processing. It must carry out quality assurance test to ensure that the products produced conform to specified standards as to the adequacy of effect of processes applied and the keeping quality of manufactured products. A good example is the phosphatase test used on pasteurised milk and the acidity development test done on U.H.T milk.

v) During marketing of processed products. Public Health authorities are employed by law to check the quality of food stuffs sold for public consumption and may impound substandard or contaminated foodstuffs including possible prosecution of culprits. This is done in order to protect the interest of the milk consuming public.

Techniques Used in Milk Testing and Quality Control

Milk Sampling

Accurate sampling is the first pre-requisite for fair and just quality control system. Liquid milk in cans and bulk tanks should be thoroughly mixed to disperse the milk fat before a milk sample is taken for any chemical control tests. Representative samples of packed products must be taken for any investigation on quality. Plungers and dippers me used in sampling milk from milk cans.

Sampling Milk for Bacteriological Testing

Sampling milk for bacteriological tests require a lot of care. Dippers used must have been sterilised in an autoclave or pressure cooker for at least 15mm at 120° C before hand in order not to contaminate the sample. On the spot sterilisation may be employed using 70% Alcohol swab and flaming or scaling in hot steam or boiling water for 1 minute.

Preservation of Sample

Milk samples for chemical tests: Milk samples for butterfat testing may be preserved with chemicals like Potassium dichromate (1 Tablet or ½ ml 14% solution in a ¼ litre sample bottle is adequate.) Milk samples that have been kept cooling a refrigerator or ice-box must first be warmed in water bath at 40 °C, cooled to 20°C, mixed and a sample then taken for butterfat determination. Other preservative chemicals include Sodium azid at the rate of 0.08% and Bronopol (2-bromo-2-nitro-1,3-propanediol) used at the rate of 0.02%.

If the laboratory cannot start work on a sample immediately after sampling, the sample must be cooled to near freezing point quickly and be kept cool till the work can start. If samples are to be taken in the field e.g. at a milk cooling centre, ice boxes with ice pecks are useful.

Labelling and Records Keeping

Samples must be clearly labelled with name of farmer or code number and records of dates, and places included in standard data sheets. Good records must be kept neat and in a dry place. It is desirable that milk producers should see their milk being tested, and the records should be made available to them if they so require.

Common Testing of Milk

Organoleptic Tests

The organoleptic test permits rapid segregation of poor quality milk at the milk receiving platform. No equipment is required, but the milk grader must have good sense of sight, smell and taste. The result of the test is obtained instantly, and the cost of the test are low. Milk which cannot be adequately judged organoleptically must be subjected to other more sensitive and objective tests.

Procedure:

- Open a can of milk.
- Immediately smell the milk.
- Observe the appearance of the milk.

- If still unable to make a clear judgement, taste the milk, but do not swallow it. Spit the milk sample into a bucket provided for that purpose or into a drain basin, flush with water.
- Look at the can lid and the milk can to check cleanliness.

Abnormal smell and taste may be caused by:

- Atmospheric taint (e.g. barny/cowy odour).
- Physiological taints (hormonal imbalance, cows in late lactation-spontaneous rancidity).
- Bacterial taints.
- Chemical taints or discolouring.
- Advanced acidification (pH < 6.4).

Clot on Boiling (C.O.B) Test

The test is quick and simple. It is one of the old tests for too acid milk (pH<5.8) or abnormal milk (e.g. colostral or mastitis milk). If a milk sample fails in the test, the milk must contain many acid or rennet producing microrganisms or the milk has an abnormal high percentage of proteins like colostral milk. Such milk cannot stand the heat treatment in milk processing and must therefore be rejected.

Procedure: Boil a small amount of milk in a spoon, test tube or other suitable container. If there is clotting, coagulation or precipitation, the milk has failed the test. Heavy contamination in freshly drawn milk cannot be detected, when the acidity is below 0.20-0.26% Lactic acid.

The Alcohol Test

The test is quick and simple. It is based on instability of the proteins when the levels of acid and/or rennet are increased and acted upon by the alcohol. Also increased levels of albumen (colostrum milk) and salt concentrates (mastitis) results in a positive test.

Procedure: The test is done by mixing equal amounts of milk and 68% of ethanol solution in a small bottle or test tube. (68 % Ethanol solution is prepared from 68 mls 96% (absolute) alcohol and 28 mls distilled water). If the tested milk is of good quality, there will be no coagulation, clotting or precipitation, but it is necessary to look for small lumps. The first clotting due to acid development can first be seen at 0.21-0.23% Lactic acid. For routine testing 2 mls milk is mixed with 2 mls 68% alcohol.

The Alcohol-Alizarin Test

The procedure for carrying out the test is the same as for alcohol test but this test is more informative. Alizarin is a colour indicator changing colour according to the acidity. The Alcohol Alizarin solution can be bought ready made or be prepared by adding 0.4 grammes alizarin powder to 1 litre of 61% alcohol solution.

Acidity Test

Bacteria that normally develop in raw milk produce more or less of lactic acid. In the acidity test the acid is neutralised with 0.1 N Sodium hydroxide and the amount of alkaline is measured. From this, the percentage of lactic acid can be calculated. Fresh milk contains in this test also "natural acidity" which is due to the natural ability to resist pH changes. The natural acidity of milk is 0.16-0.18%.

Apparatus:

- A porcelain dish or small conical flask
- 10 ml pipette, graduated
- 1 ml pipette
- A Burette, 0.1 ml graduations
- A glass rod for stirring the milk in the dish
- A Phenophtalein indicator solution, 0.5% in 50% Alcohol
- N Sodium hydroxide solution.

Procedure: 9 ml of the milk measured into the porcelain dish/ conical flask,1 ml Phenopthalein is added and then slowly from the burret, 0.1 N Sodium hydroxide under continuous mixing, until a faint pink colour appears. The number of mls of Sodium hydroxide solution divided by 10 expresses the percentage of lactic acid.

Resazurin Test

Resazurin test is the most widely used test for hygiene and the potential keeping quality of raw milk. Resazurin is a dye indicator. Under specified conditions Resazurin is dissolved in distilled boiled water. The Resazurin solution can later be used to test the microbial activity in a given milk sample.

Resazurin can be carried out as:

- 10 min test.
- 1 hr test.
- 3 hr test.

The 10 min Resazurin test is useful and rapid, screening test used at the milk platform. The 1 hr test and 3 hr tests provide more accurate information about the milk quality, but after a fairy long time. They are usually carried out in the laboratory.

Apparatus and reagents:

- Resazurin tablets
- Test tubes with 10 mls mark
- 1 ml pipette or dispenser for Resazurin solution.
- Water bath thermostatically controlled
- Lovibond comparator with Resazurin disc 4/9.

Procedure: The solution of Resazurin as prepared by adding one tablet to 50 mls of distilled sterile water. Rasazurin solution must not be exposed to sunlight, and it should not be used for more than eight hours because it losses strength. Mix the milk and with a sanitized dipper put 10 mls milk into a sterile test tube.

Add one ml of Resazurin solution, stopper with a sterile stopper, mix gently the dye into the milk and mark the tube before the incubation in a water bath, place the test tube in a Lovibond comparator with Resazurin disk and compare it colourimetrically with a test tube containing 10 ml milk of the same sample, but without the dye (Blank).

The Gerber Butterfat Test

The fat content of milk and cream is the most important single factor in determining the price to be paid for milk supplied by farmers in many countries.

Also, in order to calculate the correct amount of feed ration for high yielding dairy cows, it is important to know the butterfat percentage as well as well as the yield of the milk produced. Further more the butterfat percentage in the milk of individual animals must be known in many breeding programmes. Butterfat tests are also done on milk and milk products in order to make accurate adjustments of the butterfat percentage in standardised milk and milk products.

Apparatus for DF test:

- Gerber butyrameters, 0-6% or 0-8% BF
- Rubber stoppers for butyrometers
- 10.94 or 11 ml pipettes for milk

- 10 mls pippetes or dispensers for Gerber Acid
- 1 mls pippetes or dispensers for Amyl alcohol
- stands for butyrometers.

Gerber water bath Reagents:

- Gerber sulphuric acid,(1.82 g/cc)
- Amyl alcohol.

Treatment of Samples

Fresh milk at approximately 20°C should be mixed well. Samples kept cool for some days should be warmed to 40°C, mixed gently and cooled to 20°C before the testing.

Procedure: Add 10 mIs sulphuric acid to the butyrometer followed by 10.94 or 11 mls of well mixed milk. Avoid wetting of the neck of the butyrometer.

Next add 1 ml of Amyl alcohol, insert stopper and shake the butyrometer carefully until the curd dissolves and no white particles can be seen. Place the butyrometer in the water bath at 65°C and keep it there until a set is ready for centrifuging. The butyrometer must be placed in the centrifuge with the stem (scale) pointing towards the centre of the centrifuge.

Spin for 5 min. at 1100 rpm.

Remove the butyrometers from the centrifuge.

Put the butyrometers in a water bath maintained at 65°C for 3 min. before taking the reading.

(Note: When transferring the butyrometers from the centrifuge into the water bath make sure that the butyrometers are all the time held with the NECK POINTING UP).

The fat column should be read from the lowest point of the meniscus of the interface of the acid-fat to the 0-mark of the scale and read the butterfat percentage.

The butyrometers should be emptied into a special container for the very corrosive liquid of acid-milk, and the butyrometers should be washed in warm water and dried before the next use.

Appearance of the Test:

- The colour of the fat column should be straw yellow.
- The ends of the fat column should be clearly and sharply defined.

- The fat column should be free from specks and sediment.
- The water just below the fat column should be perfectly clear.
- The fat should be within the graduation.

Problems in Test Results

Curdy tests:

- Too lightly coloured or curdy fat column can be due to:
- Temperature at milk or acid or both too low.
- Acid too weak.
- Insufficient acid.
- Milk and acid not mixed thoroughly.

Charred tests:

- Darkened fat column containing black speck at the base is due to:
- Temperature of milk-acid mixture too high.
- Acid too strong.
- Milk and acid mixed too slowly.
- Too much acid used.
- Acid dropped through the milk.

The Lactometer test

Addition of water to milk can be a big problem where we have unfaithful farm workers, milk transporters and greedy milk hawkers. A few farmers may also fall victim of this illegal practice. Any buyer of milk should therefore assure himself/herself that the milk he/she purchases is wholesome and has not been adulterated. Milk has a specific gravity. When its adultered with water or other materials arc added or both misdeeds are committed, the density of milk change from its normal value to abnormal.

The lactometer test is designed to detect the change in density of such adulterated milk. Carried out together with the Gerber butterfat test, it enables the milk processor to calculate the milk total solids (% TS) and solids not fat (SNF). In normal milk SNF should not be below 8.5% according to Kenya Standards(KBS No 05-10:-1976).

Procedure: Mix the milk sample gently and pour it gently into a measuring cylinder (300-500). Let the Lactometer sink slowly into the milk. Read and record the last Lactometer degree (°L) just above the surface of the milk. If the temperature of the milk is different from the calibration temperature (Calibration temperature may be=20 0C) of the lactometer, calculate the temperature correction. For each °C above the calibration temperature add 0.2°L; for each °C below calibration temperature subtract 0.2 °L from the recorded lactometer reading.

For the calculations, use lactometer degrees, and for the conversion to density write 1.0 in front of the true lactometer reading,i.e. 1.030 g/ml. Clever people may try to adulterate milk in such a way that the lactometer cannot show the adulteration. But look to see if there is an unusual sediment from the milk at the bottom of the milk can and taste to find out if the milk is too sweet or salty to be normal. Samples of milk from individual cows often have lactometer reading outside the range of average milk, while samples of milk from herds should have readings hear the average milk, but wrong feeding, may result in low readings. Kenyan standards expects milk to have specific gravity of 1.026-1.032 g/ml which implies a Lactometer reading range of 26.0-32.0 °L. If the reading is consistently lower than expected and the milk supplier disputes any wrong doing arrange to take a genuine sample from the supplier (i.e. inspect milk right from source).

Freezing Point Determination

The freezing point of milk is regarded to be the most constant of all measurable properties of milk. A small adulteration of milk with water will cause a detectable elevation of the freezing point of milk from its normal values of-0.54°C. Since the test is accurate and sensitive to added water in milk, it is used to detect whether milk is of normal composition and adulterated.

Inhibitor Test

Milk collected from producers may contain drugs and/or pesticides residues. These when present in significant amounts in milk may inhibit the growth of lactic acid bacteria used in the manufacture of fermented milk such as Mala, cheese and Yoghurt, besides being a health hazard.

Principle of the Method: The suspected milk sample is subjected to a fermentation test with starter culture and the acidity checked after three (3) hours. The values of the titratable acidity obtained is compared with titratable acidity of a similarly treated sample which is free from any inhibitory substances.

Materials:

- test tubes
- Starter culture
- lml pipette
- water bath
- material for determination of titratable acidity.

Procedure: Three test tubes are filled with 10 ml of sample to be tested and three test tubes filled with normal milk.

All tubes are heated to 90 0C by putting them in boiling water for 3-5 minutes.

After cooling to optimum temperature of the starter culture (30,37, or 42°C), 1 ml of starter culture is added to each test tube, mixed and incubated for 3 hours.

After each hour, one test tube is from the test sample and the control sample is determined.

Assessment of results:

- If acid production in suspected sample is the same as the normal sample, then the suspect sample does not contain any inhibitory substances;
- If acid production as suspect sample is less than in the normal milk sample, then, the suspect sample contains antibiotics or other inhibitory substances.

Quality Control of Pasteurised Milk

When milk is pasteurised at 63°C for 30 min in batch pasteuriser or 72°C for 15 seconds in heat exchanger, continuous flow pasteurisers, All Pathogenic Bacteria Are Destroyed, there by rendering milk safe for human consumption. Simultaneously various enzymes present in milk, and which might affect its flavour, are destroyed.

In order to determine whether or not milk has been adequately pasteurised, one of the enzymes normally present in milk phosphatase, is measured. A negative phosphatase result indicates that the enzyme and any pathogenic bacteria have been destroyed during pasteursation. If it is positive, it means the pasteurisation process was inadequate and the milk may not be safe for human consumption and will have a short shelf life.

- Test tubes
- 5 mls pipettes
- 1 ml pipettes
- 100 ml volumetric flask
- 500 ml volumetric flask
- water bath at 37°C.

Note: All glassware must be rinsed, cleaned, rinsed in chromic acid solution and boiled in water for 30 min.

Reagent

Buffer Solution: Is mixed by 0.75g anhydrous sodium carbonate and 1.75g Sodium bicarbonate in 500 ml distilled water.

Buffer-substrate Solution: Place 0.15 g of di-sodium paranitrophenylphosphate (the substrate) into a clean 100ml measuring cylinder.

Add the buffer solution to make to 100 ml mark.

Store this buffer-substrate solution in a refrigerator and protected against light. It should not be used after one week. Prepare a fresh stock.

Procedure: Pipette 5mls buffer-substrate solution into a test tube, stopper and warm the solution in the water bath at 37°C. Add to the test tube 1ml of the milk to be tested, stopper and mix well and place in water bath at 37°C. Prepare a blank sample from boiled milk of the same type as that undergoing the test. Incubate both the test samples and the blank sample at 37°C for 2hrs. After incubation, remove the tubes and mix them thoroughly.

Place one sample against the blank in a Lovibond comparator" All Purposes" using A.P.T.W. disc and rotate the disc until the colour of the test sample is matched and read the disc number.

Interpretation:

Disc Reading after 2 hrs incubation at 37°C	*Remarks*
0-10	Properly pasteurised
10-18	Slightly under pasteurised
18-42	UNDER PASTEURISED
> 42	NOT PASTEURISED

Common Adulterants/Contaminants in Food and Simple Screening Tests for their Detection

Adulteration in food is normally present in its most crude form, prohibited substances are either added or partly or wholly substituted. In India normally the contamination/adulteration in food is done either for financial gain or due to carelessness and lack in proper hygienic condition of processing, storing, transportation and marketing. This ultimately results that the consumer is either cheated or often become victim of diseases.

Such types of adulteration are quite common in developing countries or backward countries. However, adequate precautions taken by the consumer at the time of purchase of such produce can make him alert to avoid procurement of such food. It is equally important for the consumer to know the common adulterants and their effect on health.

2

Factors Influencing the Quantity and Quality of Milk

The Physiology of Mammary Glands During the Dry Period and the Relationship to Infection

The mammary gland of the dairy cow requires a non-lactating period prior to an impending parturition in order to optimize milk production in the subsequent lactation. This non-lactating period is commonly referred to as the "dry period" and is required for reasons other than nutrition. While dry periods can be of any length, there is general agreement that an approximate 60-day non-lactating period is required and that dry periods of less than 40-50 days result in less than optimal milk yields.

Mastitis research workers have observed for many years that the dry period is critically related to the dynamics of intramammary infection (IMI) within a dairy herd. New IMI during the dry period contributes in a major way to the increasing number of infected quarters that occurs with each successive lactation. This critical relationship of the dry period to bovine mastitis resulted in the development of "dry cow therapy" as a necessary component of mastitis control. While it can be emphatically stated that dry cow therapy, when practiced in conjunction with post-milking teat disinfection, constitutes one of the most significant advances yet achieved in mastitis control, there is adequate reason to continue to investigate the relationship between intramammary infection and the dry period.

Firstly, as presently formulated and administered, dry cow therapy products are not equally effective against all pathogens. Secondly, the susceptibility of the mammary gland to new IMI by a particular pathogen may vary from one portion of the dry period to another, and thus, while new IMI by Streptococcus agalactiae or Staphylococcus aureus most likely occurs during the early dry period, new IMI by the environmental streptococci or coliform bacteria may occur frequently during the immediate prepartum period. A third reason for continued investigation is that the dry period, and in particular the early dry period, affords a unique opportunity to manipulate factors of natural resistance which are present in mammary secretion and which normally elevate markedly during the process of involution.

Physiological Function During the Dry Period

The physiology of the mammary gland during the dry period differs markedly from that during lactation. During lactation the primary function of the mammary gland is one of constant and continuous synthesis and secretion of milk. However, during the dry period the gland can be characterized as progressing through three distinct functional states, which are 1) the period of active involution; 2) the period of steady state involution; and 3) the period of lactogenesis and colostrogenesis.

The period of active involution begins with the cessation of regular milking and is probably completed by 30 days into the dry period. The period of steady state involution does not have definitive points by beginning and ending but represents the period of time the gland is maintained in the fully involuted state.

The length of this functional state will increase or decrease proportional to the length of the dry period and is in contrast to the other two functional periods which are controlled by hormonal or management events. The third functional state is that of lactogenesis-colostrogenesis and is characterized by regeneration and differentiation of secretory epithelial cells, selective transport and accumulation of immunoglobulin and the onset of copious secretion. The active period of lactogenesis and colostrogenesis most likely begins 15-20 days prepartum.

Thus, at both the beginning and at the end of the dry period, the mammary gland is undergoing transition either from, or to, a state of active synthesis of copious quantities of secretion. The bulk of research evidence available would suggest that these periods of transition from one functional state to another, occurring in the absence of regular twice daily milking, are periods of heightened susceptibility to new IMI.

Active Involution-The Early Dry Period

The dry period is initiated by cessation of milking at the end of lactation. However, the mammary gland continues to synthesize and secrete milk which accumulates in the gland. Studies conducted with cows producing 9-10 kg of milk per day at drying off showed that mammary glands accumulated 75-80% of their daily yield at the time of drying off and that maximum accumulation of fluid occurred between 2 and 3 days post drying off. Fluid volume continues to decrease through at least day 16 of the dry period and probably through 30 days of involution.

The increase in intra-alveolar pressure as a result of milk accumulation is thought to trigger the events of active involution. An early event in involution is the appearance of lysosomes in the secretory epithelial cells. The lysosomes are thought to participate in autophagocytosis of secretory cell constituents. As the cells are degraded there is a loss of cell to cell contact as well as a loss of contact with the basement membrane. Heterophagocytosis of the degraded epithelial cells and the accumulated fat and casein is thought to be a function of the macrophages which enter the mammary tissue and secretion in large numbers during active involution.

As a result of these early events in involution, the composition of the secretion changed markedly. The rate of synthesis of the major milk constituents, i.e. fat, casein, lactose, citrate, â-lactoglobulin and á-lactalbumin decreases markedly by 3 to 4 days into involution. The actual concentration of these major constituents in the secretion of the involuting gland varies. The concentration of fat, casein, â-lactoglobulin and á-lactalbumin appears not to undergo major or consistent change. In contrast, the concentration of lactose and citrate, major regulators of osmolarity of mammary secretion, decrease markedly with involution but not prior to 3 days involution.

Proteins of mammary secretions whose origin is thought to be blood serum, are commonly present in whey of normal milk. The concentration of these proteins changes very little through 3 days involution, but by 7 days involution, the concentration of serum proteins in mammary secretion is significantly elevated. Thus, the marked increase in serum albumin and IgG immunoglobulin that occurs between 3 and 7 days if involution is consistent with the period of major reduction in mammary fluid volume. The concentration of all immunoglobulin classes, i.e. IgG_1,, IgG_2, IgA and IgM increases substantially by 7 days involution and there is a transitory increase in the selective transfer of IgG_1 which occurs at 2 to 3 days involution. The magnitude of increase in selective transfer is far less than that observed during colostrogenesis, and the transitory nature of the increase may suggest that receptor sites required for selective transfer are either temporarily synthesized in greater numbers or made accessible as a result of the early involution events occurring in the secretory epithelial cell.

The increase in serum albumin (BSA) probably reflects an increased permeability of the secretory epithelial cell barrier to passive diffusion of serum proteins. The concentration of BSA during involution never approaches the level found in blood serum or the concentration found in mammary secretion during acute inflammation. These findings suggest that the permeability barriers are not totally destroyed and that the mammary gland maintains a degree of control even though degenerative processes are occurring within the tissue.

A characteristic compositional change associated with involution is the dramatic increase in concentration of the iron biding protein lactoferrin (Lf). Lactoferrin is not specific to milk as it is present in the secretions of most mucosal epithelial surfaces throughout the body. Lactoferrin is a major constituent of the secondary granules of the polymorphonuclear neutrophil (PMN) and has been found in normal human blood serum at 0.8 ìg/ml. The major site of synthesis of the Lf found in bovine mammary secretions is thought to be the secretory epithelial cell with the PMN making only a minor contribution.

In normal bovine milk Lf is a minor whey protein and is found at 0.1-0.3 mg/ml. This is in marked contrast to the concentration in human milk (3-5 mg/ml) where it would appear to be the major whey protein. The concentration of Lf increases markedly in mammary secretion during active involution of the mammary gland in both the cow and the human.

We have previously postulated that the increased numbers of lysosomes observed in secretory cells undergoing active involution is the source of the Lf observed in secretions of involuting mammary glands. The concentration of Lf is in the order of 0.4 mg/ml at drying off, increases to 1.4 mg/ml at 3 days involution, 8.5 mg/ml at 7 days involution and is generally found at 20-30 mg/ml by 30 days into the dry period. Thus, Lf is a major protein in the secretion of the non-lactating mammary gland.

Lactoferrin is thought to participate in the non-specific defence of the involuting mammary gland and has been reported to be of particular importance with regard to coliform infection. Lactoferrin is bacteriostatic by virtue of its ability to bind iron with great affinity. Thus, invading bacteria are forced to compete with Lf for iron which is required for normal growth of most pathogens. The possible exception is the streptococcal species which apparently have an extremely low iron requirement. However, there are also indications that Lf or peptides derived from LF can be bactericidal.

With regard to coliform infection, the bacteriostatic properties of Lf are reversed in the presence of citrate, a normal constituent of milk. Citrate, like Lf, is a chelator of iron and coliform bacteria that can utilize citrate-bound iron. The molar ratio of citrate to Lf determines the degree of coliform growth inhibition in vitro. As the molar ratio decreased, inhibition of coliform growth increased.

During involution the molar ratio of citrate to Lf decreases significantly. Examples of changes in the molar ratios are 2,356, 1,197, 581 and 49 at 0, 2, 3 and 7 days involution, respectively. The coliform growth inhibitory properties of whey collected during the first seven days of involution increased as the citrate to Lf molar ratio decreased. Thus, it would appear that the citrate to Lf molar ratio of bovine mammary secretions is inversely related to the ability of that secretion to inhibit coliform growth. In addition, the molar ratio of citrate to Lf serves as a sensitive marker system for the processes of involution and lactogenesis.

While there are numerous reports in the literature supporting the concept of Lf being a mediator of non-specific resistance to infection, there is a rapidly growing body of evidence suggesting that Lf may play an important role in the modulation and control of the functional processes of macrophages, lymphocytes and the PMN. Lactoferrin has been shown to interact specifically with receptors on the surface of the macrophage and the lymphocyte. As previously mentioned, Lf is a normal constituent of the PMN secondary granule and limited binding of Lf to the PMN surface has been reported. Lactoferrin may have an autoregulatory role to retain PMN at inflammatory sites and thus amplifies the inflammatory response.

Lactoferrin has been reported to be involved in the control of granulopoesis by inhibiting the production of granulocyte colony stimulating factor by the macrophage. There are reports that Lf can modulate lymphocyte function, and these reports suggest that Lf suppresses the response of these cells to antigen or mitogen. In addition, the iron binding proteins, namely Lf, transferrin and ferritin, play a role in directing the migration of cells with appropriate receptors and Lf may play a role in directing migration of mononuclear cells into acute inflammatory lesions and certain areas of lymphoid tissue. That Lf receptors have been demonstrated on the surface of the macrophage and lymphocyte suggests that Lf may direct or control the influx of these cell types into mammary tissue during involution, and there is reason to speculate that once in the tissue of the involuting mammary gland, the high concentration of Lf may influence or modulate their functional activity.

The major cell types present in the secretion of the involuting bovine mammary gland are the macrophage, lymphocyte and PMN. Data consistently show that there are virtually no free secretory epithelial cells in the secretion of the involuted gland. The concentration of cells increases during the first week of involution, but prior to three days involution, there would appear to be very little increase over the level present at drying off. This is consistent with the lack of change in other parameters during the first three days of involution.

Mammary quarters infected at drying off or quarters that become infected have higher cell counts during involution than do uninfected quarters and the PMN would appear to be the predominant cell type present in the secretion during the first 7 days of involution. By 14 days involution, the macrophage is the dominant cell type present, but lymphocyte numbers are also increased. Lymphocyte numbers continue to increase between 14 and approximately 30 days involution, and this is in contrast to the marked decline in PMN numbers and a lesser decline in macrophage numbers over this period.

The lymphocyte would appear to be the major cell type present in mammary secretion at 30 days involution. The functional role of the phagocytic cells during early involution is generally thought to be that of phagocytosis of degenerated secretory epithelial cells, fat and casein.

Presumably the macrophage plays the major role in removing the large quantities of accumulated fat and cellular debris. The marked decline in fluid volume and the reduction in the quantity of fat and casein between 3 and 7 days involution, would suggest that during this period virtually all phagocytic cells would be functionally committed to clearing the gland of these accumulated secretory products.

The functional role of lymphocytes present in the secretion of the involuting gland is virtually unknown. The lymphocytes would appear to consist of both B and T cell populations in ratios comparable to blood. A consistent finding has been that lymphocytes from bovine mammary glands are in general less responsive to mitogens than lymphocytes isolated from blood.

There is some suggestion that many of these lymphocytes may have already undergone stimulation as a result of their presence in the secretion. On the other hand, Lf reduced the responsiveness of blood lymphocytes to the mitogen concanavalin A. The high concentration of lymphocytes in the involuting gland has been interpreted to suggest that the involuting or involuted gland would be more responsive to local immunization, and data have been published which support this hypothesis.

The Relationship of Active Involution to New IMI

There is little doubt that the rate of new IMI increases markedly during the process of active involution. The changes in mammary function and composition of the secretion of the gland could be interpreted to relate both positively and negatively to new IMI. Factors which may favour new infection during the early dry period are the following.

1) A large volume of milk is accumulated and leakage of milk from teats is often observed.
2) The contents of the gland are no longer being removed at regular intervals.
3) Teat end disinfection is stopped.
4) There is only minimal change in the composition of fluid through the first 3 days of involution. Thus, the factors of natural defence are only marginally elevated over normal milk levels during this potentially vulnerable period.
5) The phagocytic cells, a primary means of eliminating bacteria which gain entrance to the mammary gland, are occupied in the phagocytosis of degenerated epithelial cells, fat and casein.
6) There is reason to suspect that there are factors (other than fat and casein) present in the secretion of the involuting gland that may reduce the phagocytic capabilities of the macrophage and PMN and reduce the responsiveness of lymphocytes to antigen stimulation.

7) There would appear to be secretion components which are either increased in concentration or are produced during the process of involution, which can act as growth stimulants for certain bacteria.

From a positive point of view, the elevation in phagocytic cells, lymphocytes, immunoglobulins and Lf, could all be interpreted as changes occurring during involution that should render the gland more resistant to new IMI. However, it is likely that these changes occur too slowly and that their antibacterial action is compromised (as previously discussed) by other normal events of involution. Indeed, treatment of glands at or near drying off with colchicine, endotoxin or the combination of the two compounds resulted in a more rapid involution, earlier elevation of resistance factors and reduced IMI during the first week of the dry period.

Steady State Involution-Relationship to New IMI

Modern methods of dairy cattle management which stress minimum calving intervals and dry periods of 60 days or less, are likely to result in the process of involution (regression of the secretory tissue to terminal ducts with minimal numbers of alveoli present) being completed just prior to the hormonally initiated process of active lactogenesis and colostrogenesis.

We would suggest that the overlapping of the two events may lead to a less than optimal hormonally mediated lactogenic response within the tissue and thus relate to the less than optimal milk yields in lactations following dry periods of under 40 days. As a consequence of dry periods of minimal length, the mammary glands of normal pregnant dairy cows may be maintained in steady state involution for a very short time period, and in reality, the gland would progress from active involution to active lactogenesis. However, in cows that are not pregnant or when dry periods are 60 days or greater mammary glands do enter into a period of steady state involution.

During this period fluid volume is maintained at levels of only a few ml, and the major constituents of milk are greatly reduced or at minimal concentration. Lactose and citrate are at concentrations of less than 5 mg/ml and.2 mg/ml, respectively. The citrate to Lf molar ratio is maintained 100 fold lower than that found during lactation. The concentration of immunoglobulins and Lf are elevated but generally at levels less than those recorded at 4 weeks involution. The predominant cell type would appear to be the lymphocyte in uninfected quarters but substantial numbers of macrophages are present. The PMN are generally found in low numbers by comparison to lymphocytes and macrophages.

The conditions within the mammary gland during steady state involution could be argued to favour the host in most relationships between host and pathogen. Indeed, the incidence of new IMI during this period has been reported to be the lowest of any functional state of the mammary gland. The degree to which the secretion of the fully involuted gland can resist the establishment of new IMI may vary with the bacterial species.

The fully involuted gland is apparently quite resistant to the establishment of coliform infections following infusion of the organisms through the streak canal but susceptible to Str. uberis and Str. agalactiae. Thus, the marked reduction in new IMI during steady state involution may relate partially to the high levels of antibacterial factors in the secretion but may also suggest a reduced rate of bacterial penetration through the streak canal. The latter is a result of a subumlike plug of keratin within the streak canal or a substantially lower level of exposure of teat ends to potential pathogens.

In support of the contention that the secretion of the fully involuted gland is highly antibacterial, it should be pointed out that spontaneous elimination of intramammary infection is a frequent event in the fully involuted gland and that acute clinical cases of mastitis are seldom seen in the fully involuted gland.

Lactogenesis-Colostrogenesis: Relationship to New IMI

The physiology of the mammary gland during the late dry period is again characterized by transition. In contrast to the early dry period the dramatic changes occurring in the mammary tissue are constructive as opposed to the destructive processes of active involution. In addition, the change are hormonally modiated. A major function of the mammary gland during this period is the formation and accumulation of colostrum and a unique feature of colostrum is the high concentration of the immunoglobulin IgG_1. The major physiological role of colostrum is that of a source of passively acquired antibody for the newborn calf.

The concentration of the major components of milk begin to increase in secretion approximately two weeks prior to parturition with marked increases in fat, casein and lactose occurring in the 5 days preceding parturition. The volume of fluid in the gland cistern increases slowly over the last two weeks of the dry period and then increases dramatically 1-3 days prepartum with the onset of copious secretion. The rapid increase in fluid volume is highly correlated with the dramatic increase in the concentration of citrate. The increasing concentration of citrate as a "harbinger" of lactogenesis or the onset of copious secretion.

During the period beginning approximately two weeks prepartum, the concentration of Lf declines to levels of 2-3 mg/ml found in colostrum. As a result of the changes in citrate and Lf, the molar ratio (citrate:Lf) increases by approximately 100 fold over the last few days prepartum. This dramatic increase in the citrate to Lf molar ratio suggests a loss of the antimicrobial properties of Lf and particularly in regard to coliform bacteria.

The concentration of immunoglobulin and in particular the IgG_1 begins to increase 2-3 weeks prepartum. Maximum concentration of all Ig classes is achieved approximately 5-10 days prepartum. The concentration falls as fluid accumulates in the gland with the onset of copious secretion. Approximate concentrations of IgG_1, IgG_2, IgA and IgM in colostrum are 60 mg/ml, 2.5 mg/ml, 4.5 mg/ml and 6.0 mg/ml, respectively.

Number of cells in secretion of uninfected quarters gradually decline over the 2 week prepartum period. The macrophage would again appear to be the predominant cell type. This is in contrast to the fully involuted gland where the lymphocyte is the most numerous cell. A most striking feature of the changing cellular content is the approximate 100 fold decline in lymphocyte numbers during the period 2 weeks prepartum. That the decline in lymphocyte numbers occurs coincidentally with the decrease in Lf concentration is of interest in light of the hypothesis that iron binding proteins may be involved in directing lymphocyte traffic into tissues.

The phagocytic ability of macrophages and PMN appear to be inhibited by the secretion of the prepartum gland or their function is reduced as a result of a general immunosuppression known to exist around parturition.. Macrophages isolated from colostrum are observed to contain numerous fat globules and may be one factor limiting phagocytic potential. However, research has suggested that inhibitors other than fat and casein may be present in the secretion of the prepartum mammary gland. The period immediately prepartum is known to be a period of increased susceptibility to new IMI caused by the environmental pathogens and dry cow therapy products as currently formulated and administered afford little or no protection to the gland during the latter portion of the dry period.

Management of Dairy Cows

Economic Characters in Dairy Cattle

The various economic characters in Dairy Cattle management are:

1. Lactation yield
2. Lactation period
3. Persistency of yield
4. Age at first calving
5. Service period
6. Dry period
7. Inter calving period
8. Reproductive efficiency
9. Efficiency of feed utilization
10. Disease resistance.

Lactation yield: The lactation yield in a lactation period is known as lactation yield. 'Thelactation yield in Indian breeds is very low compared to exotic breeds. This is dependent on no.. of calving, frequency of milking, persistency of yield (Normally in dairy cattle 30-40 % increase in milk production from first lactation to maturity is observed. After 3 or 4 lactation the production starts declining. For comparison of milk yield of different breeds and animals the mil yield should be converted into fat corrected milk (FCM).

4% FCM = 0.4 total milk + 15 total fat. After parturition the milk yield per day will be increased and reaches peak within 2-4 weeks after calving. This yield is known as peak yield. The maintenance of peak yield for more time is importance for better milk production. The lactation period in Indian breeds is low and so the production is also less and conversion.

Lactation period :The length of milk producing period after calving is known as lactation period. The optimum lactation period is 305 days. The milk production will Breeding of dairy animals and farm records 33 be less, if this period is shortened". Indian breeds will have less lactation period, but in some breeds this period is more with very little milk production Persistency of Milk Yield :During lactation period the animal reaches maximum milk yield per day with in 2-4 weeks which is called peak yield.

For high level of lactation yield, this peak yield should be maintained for longer period as far as possible, The maintenance of peak yield for long period is known as persistency, slow decrease in dairy milk yield after reaching peak yield in necessary. High persistency is necessary to maintain high level of milk production Age at first calving: The age o the animal at first calving is very important for high life time production. The desirable age at first calving in Indian breeds is 3 years, 2 years in cross breed cattle and 3 1/2 years in Buffaloes. Prolonged age at first calving will have high production in the first lactation) but the life time production will be decreased due to less no of calvings. If the age at first calving is below optimum, the calves born are weak, difficulty in calving and less milk production in first lactation.

Service period : It is the period between-date of calving and date of successful conception. The optimum service period helps the animal to recover from the stress of calving and also to get back the reproductive organs back to normal For cattle the optimum service period is 60-90 days. If the service period is too prolonged the calving interval prolonged, less no. of calvings will be obtained in her lifetime and ultimately less fife time production.' If the service period is too short, the animal will become weak and persistency of milk production is poor due to immediate pregnancy.

Dry Period: It is the period from the date of drying (stop of milk production) to next calving When the animal is in pregnancy, before next calving. The animal should be given rest _period to compensate for growth of foctus. A minimum of 2-234 Dairy Animal Management½ months dry period should be allowed) If the dry period is not given or too low dry period, the animals suffer from stress and in next lactation, the milk production drops substantially and also it gives weak calves. On the other hand if the dry period given is too high, it may not have that much effect on increasing milk yield in the next lactation, but it decrease the production in the present lactation.

Inter calving period : This is the-period between two successive calvings. It is more, profitable to have one calf yearly in cattle and at least one calf for every 15 months in buffaloes. If the calving interval is more, the total no. of carvings in he life time will be decreased and also total life production of milk. decrease. Reproductive Efficiency : The reproductive efficiency means the more number of calves during lifetime, so that total life time production is increased, The reproduction or breeding efficiency is determined by the combined effect of hereditary and environment. Several measures of breeding efficiency like number of services per conception, calving interval, days from first breeding to conception are useful.) Reproductive efficiency has generally a low heritability value indicating that most of the variations in this trait is due to non genetic factors. In adverse environmental conditions, the poor milk producing animals may not be much effected compared to high effect in high milk yield.

Efficiency of Feed Utilization and Conversion into Milk : The animal should utilize the feed efficiently to convert into the milk. Disease Resistance : Indian breeds are more resistant to majority of disease compared to exotic cattle. Cross breeding helps to get this character.

System of Breeding

Breeding is defined as the crossing of the male and the female parents to get the off spring for the characters desired.

Housing of Dairy Cattle

Selection of Site for Dairy Farm: The points which should be considered before the erection of dairy buildings are as follows :

Topography : A dairy building should be at a higher elevation than the surrounding ground to offer a good slope for rainfall and drainage for the wastes of the dairy to avoids tagnation within. A levelled area requires less site preparation and thus lesser cost of building. Low lands and depression and proximity to places of badod our should be avoided.

Soil Type: Fertile soil should be spared for cultivation. Foundation soil as. far as possible should not-too dehydrated or desiccated. Such a soil is susceptible to considerable swelling during rainy season and exhibit numerous cracks and fissures.

Exposure to the sun and protection from wind : A dairy building should be located to a maximum exposure to the sun in the north and minimum exposure to the sun in the south and protection from prevailing strong wind currents whether hot or cold. Buildings should be placed so that direct sunlight can reach the platforms, gutter and mangers; in the cattle shed. As far as possible, the long axis of the dairy barns should-be set in north south direction to have the maximum benefit of the sun.

Accessibility: Easy accessibility to the buildings is always desirable. Situation of a cattle shed by the side of the main road preferably a distance of about 100 meters should be aimed at.

Durability and Attractiveness : It is always attractive when the buildings open upto a scenic view and add to the gradeur of the scenery. Along with this, durability of the structure is obviously an important criteria in building a dairy.

Water Supply : Abundant supply of fresh, clean and soft water should be available at a cheap rate.

Surrounding : Areas infested with wild animals and 'daoits; should be avoided. Narrow gates, high manger curbs, loose hinges, protruding nails, smooth finished floor in the areas where the cows move and other such hazards-should be eliminated.

Labour : Honest, economic and regular supply of labour be available.

Marketing : Dairy buildings should only be in those areas from where the owner can sell his products profitability and regularly. He should be in a position to satisfy the needs of the farm within no time and at a reasonable price.

Electricity : Electricity is the most important sanitary method of lighting a dairy. Since a modern dairy always handles electric equipments which are also economical, it is desirable to have an adequate supply of electricity.

Facilities, labour and food : Cattle yards should be so constructed and situated in relation to feed storages, hay stacks, silo and manure pits as to effect the most efficient utilization of labour. Sufficient space per cow well arranged feeding mangers and resting areas contribute not only to greater milk yield of cows and make the work of the operator easier but also minimise feed expenses. The relative position of the feed stores should be quite adjacent to the cattle barn. Noteworthy features of feed stores are given below.

a) Feed storages should be located at hand near the centre of the cow barn.

b) Milk house should be located almost at the centre of the barn.

c) Centre cross-alley should be well designed with references to feed storage, the 'stall areas and the milk house.

Systems of Housing-Loose Housing System

Conventional Dairy Barn

The most widely prevalent practice in this country is to tie the cows with rope on a Katcha floor except some organised dairy farms belonging to government, co-operatives or military where proper housing facilities; exist. It is quite easy to understand that unless cattle are provided wit good housing facilities, the animals will move too far in or out of the standing space, defeating all round and even causing trampling an wasting of feed by stepping into the managers. The animals will be exposed to extreme weather conditions all leading to bad health and lower production.

Dairy cattle may be successfully housed a wide variety of condition, ranging from close confinement to little restrictions except at milking time.

1. The loose housing barn in combination with some type of milking barn or parlour.
2. The conventional diary barn.

Loose Housing System

Loose housing may be defined as a system where animals are kept loose except milking and at the time of treatment. The system is most economical. Some features of loose housing system are as follows.

1. Cost of construction is significantly lower than conventional type.
2. It is possible to make further expansion without change
3. Facilitate easy detection of animal in heat.
4. Animals feel free and therefore, proves more profitable with even minimum grazing
5. Animals get optimum excise which is extremely important for better health production.
6. Over all better management can be rendered.

Sanitation in Dairy Farm

Sanitation is necessary in the dairy farm houses for eliminations of allmicro organism that are capable of causing disease in the animals. The presence of organisms in the animal shed contaminates the milk produced thus reducing its self life, milk produced in an unclean environment is likely to transmit diseases which affect human health:

Dry floorings keeps the houses dry and protects from foot injury. Similarly the presence of flies and other insects in the dairy farm area are not only, disturbs the animals but also spreads deadly diseases to the animals egg. Babesiosis, Theileriosis.

Cleaning Of Animal Sheds: The easy and quick method of cleaning animal house is with liberal use of tap water, proper lifting and disposes all of dung and used straw bedding, providing drainage, to the animal house for complete removal of liquid waste and urine. The daily removal of feed and fodder left over in the manger, reduces the fly nuisance. Periodical cleaning of water through eliminates the growth of algae, bacterial and viral contamination and thus keeps the animal healthy.

Sanitizers : Sunlight is the most potent and powerful sanitizer which destroy most of the disease producing organism. Disinfection of animal sheds means making these free from disease producing bacteria and is mainly-carried out by sprinkling chemical agents such as bleaching powder, Iodine and lodoph or, sodium carbonate, Washing soda, Slaked Lime (Calcium hydroxide), Quick Lime (Calcium oxide) and phenol.

Bleaching Powder: This is also called calcium hypo chloride. It contains upto 39 % available chlorine which has high disinfecting activity.

Iodine and lodophor :-This is commercially available as lodophores and contains between 1 and 2 % available Iodine which is an effective germicide.

Sodium Carbonate: A hot 4 % solution of washing soda is a powerful disinfectant against many viruses and certain bacteria.

Slaked Lime And Quick Lime : White washing with these agents makes the walls of the sheds and the water troughs free from bacteria.

Phenol : Phenol or carbolic acid is a very disinfectant which destroy bacteria as well as fungus.

Insecticide : Insecticides are the substances or preparations used for killing insects. In order to control flies and disease transmitting ticks, insecticides are used in dairy farms. Ticks usually hide in cracks and crevices of the walls and mangers. Smaller quantities of insecticide solutions are required for spraying. Liquid insecticides can be applied with a powerful sprayer_ hand sprayer, a sponge or brush, commonly used insecticides are BHC, DDT, Gramaxane wettable powders, malathion, surriithion, Sevin 50 % emusifying concentration solutions.

These are highly poisonous and need to be handled carefully and should not come in contact with food material, drinking, water, milk etc.

Precautions While using disinfection I Insecticide:

- Remove dung and used bedding completely.
- Avoid spilling of dung and used bedding while carrying it out.
- Avoid the use of dirty water in cleaning the sheds.
- Never put the fresh fodder over: the previous day's left over fodder in the manger.
- Prevent algae to grow in the water troughs
- Use proper concentration of disinfectant/insecticide solutions to avoid any toxic effects poisoning.
- Avoid of the mat the milking time as milk absorbs these quickly.

Procedure:

- Remove the dung from the floor and urine channel with the help of a shovel and basket (iron) and transfer it to the wheel-barrow. Remove the used bedding and leftovers from the mangers in a similar way.
- Empty the water trough and scrape its sides and bottom with the help of a floor brush.
- Wash the water trough with clean water and white wash it with the help of lime mixture once a week.
- Scrape the floor with a brush and broom and wash with water.
- Clean and disinfect the splashes of dung on the side walls, railing and stanchions.
- Remove the cobwebs periodically with the help of a wall brush.
- Sprinkly one of the available disinfecting agent in the following concentration. Bleaching powder should have more than 30% available chlorine. Phenol 1-2% solution. Washing Soda (4% solution).
- Allow adequate sunlight to enter in to the shed.
- Spray insecticides at regular intervals specially during the rainy season (Fly season).

- White wash the walls periodically by mixing insecticides in it to eliminate ticks and mites living in cracks and crevices.

Care and Management of Calf

We must give good! feeding and management for the calves so that they develop well and,; useful for replacement stock. The feeding and care of the calf begins~ before its birth The dam should be dired 6-8 weeks before expected calving and should be fed well. Underfed animals will give weak and smog calves.

Early Management:

1) Immediately after birth remove any mucous or phlegm from those nose and mouth.
2) Normally the cow licks the calf immediately the birth. This helps' dry off the calf and helps in stimulating breathing and circulation. When the cows does not lick or in cold climate, rub and dry the calf with a dry cloth or gunny bag. Provide artificial respiration by compression and relaxing the chest with hands
3) The Naval should be tied about 2-5 cms away from the body and cut 1cm below the ligature and apply Tr. Iodine or boric acid or any antibiotic.
4) Remove the wet bedding from the pen and keep the stall very clean and dry in condition.
5) The weight of the calf should be recorded.
6) Wash the cow's udder and teats preferably with chlorine solution and dry. Allow the calf to suckle the first milk of the mother i.e. Colostrum.
7) The calf will be standing and attempts to nurse within one hour. Otherwise help too weak calves.

Feeding of Calves :

1) Feed colostrum i.e. the first milk of the cow for the first 3 days. The colostrum is thick and viscous. It contains higher proportions of Vit A and proteins. The proteins are immune globulin which gives protection against many diseases. colostrums contains antitrypsin which avoid digestion of immunoglobulins in the stomach and is absorbed as it is.
2) Whole milk should be given after 3 days it is better to teach to, drink the milk from the pail or bucket. Feed twice a day which should be warmed to body temperature. For weak calves feed thrice a day.

3) The limit of liquid milk feeding is 10 % of it's body weight with a maximum of 5-6 litres per day and continue liquid milk feeding for 6.10 weeks. Over feeding causes 'Calf Scours'.
4) The milk replaces can be given to replace whole milk.
5) Give calf starter after one month of age.
6) Provide good quality green fodder and hay from 4th month afterwards.
7) Feeding of antibiotics to calves improves appetite, increases growth rate and prevents calf scours. E.g. aureomycin, Terramycin etc.

Other Management Practices:

1) Identity the calf by tattooing in the ear at birth, and branding after one year.
2) Dehorn the calf within 7-10 days after birth with red hot Iron or caustic potash stick or electrical method.
3) Deworm the calf regularly to remove worms using deworming drugs. Deworm at 30 days interval.
4) Fresh water should be given from 2nd week onwards.
5) House the calves in individual calf pens for 3 months afterwards in groups. After six months males and females calves should be housed separately.
6) Weigh the calves at weekly interval upto 6 months arid at monthly interval afterwards to know the growth rate.
7. Mortality in calves is more in first month due to pnuemonia. Diarrhea (calf scous) and worms. House them under warm condition, clean condition to avoid above condition.
8. Extra teats beyond 4 should be removed at 1-2 months of age.
9. 8-9 weeks of age, males should be castrated.
10. Keep the body clean and dry to avoid fungal infection.
11. Mineral-blocks should be provided, so that the calves lick and no changes for mineral deficiency.
12. Wean the calf from the mother and feed through pail feeding system.

Care and Management of Heifer

Better Care and Management of heifer will give high quality replacement stock to the dairy farm. The following care and Management practices are recommended for a heifer.

1. Feed the heifer sufficiently to produce normal growth. During the early stage relatively more protein than energy is needed. Most heifers grow well if excellent hay is given as much they can eat. The amount of growth depends upon the quality of forage fed.
2. The heifers should be provided with a dry shelter free from drafts. A loose housing system with a shelter open to one side is sufficient.
3. The size rather than the age of a dairy heifer at breeding time is important. Breeding under sized animals is never profitable. They may be stunted or slow to reach maximum size. Small heifer are more likely to have difficulty in calving. Though the heifer that is bred to calve at an older age yields higher milk yield in the first lactation, the total milk produced by such a cow will be less when compared to the heifers that freshens at an early age. Usually the heifer is bred to freshen at 24-30 months of age.
4. The heifer should be growing and in good flesh at calving time. This is necessary so that she can produce milk at the most profitable level.
5. Place the heifer in a separate shed about 6-8 weeks before she is due to calve.
6. Feed 2-3 kgs of concentrate daily and all the forage she eats.
7. Before calving let the heifer becomes accustomed to handling and to the procedures used in the milking herd. Always handle her gently and with kindness.
8. Maintenance of health among heifers is very important for proper growth. The health among the heifers is maintained by hygienie housing, water balanced feeding and taking necessary preventive steps against common diseases.
9. Periodically the heifers in the herd should be checked for their proper growth and other progress. Animals lagging behind below the required standards should be removed from the herd.
10. For the heifer the calving is first time and it may have difficulty in calving. So take extra care during calving.

Care and Management of Milch Animal

To get high milk during any lactation, the milch animal should be properly fed and necessary care and managemental practices should be followed.

1. Provide green succulent forage together with leguminous hay or straw to the extent of animal can consume, so that all its maintenance requirements are met with through forage only. Extra concentrate at the rate of 1 kg for every 2 to 2.5 litres of milk should be provided. Salt and mineral supplements should be given to maintain the lactation.
2. Never frighten or excite the animals. Always treat them gently and with kindness.
3. With proper feeding and care, a cow will come to heat with in 16 days of calving. Do not with hold service unnecessarily after the signs of heat are noticed in a cow. The shorter the interval between calvings, the more efficient the animal is as a milk producer. By maintaining proper records of breeding and calvings of the animals will ensure a study flow of milk through out the year.
4. Individual attention to feed each animal according to its production is a must. For this purpose maintain individual production records.
5. Keep up regularity of feeding. Concentrate mix is fed before or during milking, when as roughages after milking. This practice will avoid dust in the shed.
6. Water should be provided to drink at will or at frequent intervals. It is more beneficial, if the animal is maintained on paddy straw as sole rough age..
7. Regularity in milking is essential. Increase of milk in the udder will reduce further secretion of milk. Milking thrice is better than twice since 10-15 % more milk can be produced.
8. Rapid, continuous, dry hand milking should be practiced without undue jerking of teats, milking should be done with whole hand, but not with thumb and index finger.
9. Cows should be trained to let down milk without calf suckling. This will held to wean the calves early.
10. Loose housing with shelter during hot part of the day should be provided. The animals will get maximum exercise in loose housing system.
11. Grooming of the cows and washing of the buffaloes before milking help in clean milk production. Daily brushing will remove loose hair an dirt from the coat. Grooming will also keep the animal hide pliable.

12. Wallowing of buffaloes or water spraying on their bodies will keel the buffaloes comfortable especially in summer.
13. Common ailments should be properly detected and treated.
14. Common vices should be properly detected and care should be taken. Eg. Kicking, licking, suckling etc.
15. Provide at least 60-90 days dry period between calvings. If the dry period is not sufficient, the milk yield is subsequent lactation will be reduced.
16. Vaccinate the cows-against important diseases and also guard against insects and pests.
17. Every animal should be numbered and particulars pertaining to milk, fat %, feed taken, breeding, drying and calving dates should be recorded.
18. Check for mastitis regularly.

Care And Management of Dry and Pregnant Animal

The good care and managemental practices given to pregnant animal will give good calf and also high milk yield during the successive lactation.

1. Extra concentrate mix of 1.25 to 1.75 kgs should be provided for pregnant animal as pregnancy allowance. Feed good quality of leguminous fodder. The animal should not be not-lean-not fat' condition.
2. Provide clean drinking water and protection from thermal stress.
3. Do not allow them to mix with other animals that have aborted or that are suffering from or carriers of diseases like brucellosit.
4. Allow moderate exercise, which helps in calving normally. Do not tire them by making long distances especially on uneven surfaces.
5. Do not allow them to fight with other animals and take care that they are not chased by dogs and other animals.
6. Avoid slippery conditions, which causes the animal to fall receiving fractures, dislocation etc.
7. If accurate breeding records are available, calculate the expected date of calving. Separate it one or 2 weeks before and shifted to individual parturition pens. These pens are thoroughly cleaned and fresh bedding may be provided.

8. Feed one kg extra concentrate during last 8 weeks of gestation. Feed laxative about 3-5 days before and after calving (Wheat bran 3 kgs + 0.5 gs of Groundnut cake + 100 gms of mineral mixture of salt).
9. Symptoms of delivery may be observed i.e. swelling of external genetalia, swelling of udder, usually majority of animals will deliver without any help. If there is any difficulty, provide veterinary help.
10. After parturition external genetalia, flank should be cleaned arid protect the animal from chill and give warm wafer.
11. Placenta will normally leave the cow within 2-4 hours after calving. It not take the help of a veterinarian.
12. Take care of the animal before calving from milk fever. Give calcium supplement.
13. Some times the udder will be swollen just before calving. Remove the milk partially.
14. Take care, of the animal, if at all any abortion.
15. Provide always free access to drinking water.

Care and Management of Bulls and Bullocks

Care and Management of Bull

The maintenance of breeding bulls in good condition and suitable for breeding is highly essential requirement for the success of breeding programmes. A rising condition is better for reproduction than a falling one. Fat males may produce semen of inferior quality or they may be slow or fail at service. Breeding bull should receive plenty of exercise, will usually produce large ejaculation containing more sperms of higher activity.

A breeding bull should housed separately known as "Bull Shed" with sufficient area of floor and proper covering. It is sound practice to provide cool conditions and adequate drinking water.

A balanced rations should be fed containing adequate energy, proteins, minerals, and vitamins. Green fodder must be available both before and during breeding season.

Most of the bulls are ferocious and so control them properly using nose rings etc. It is of great importance that males should be, fed regularly and not too much at one time, and too little at another. For bulls two mating a day has been found to be openings.

Moderate exercise should be provided to keep the breeding bull in active and non fatty conditions. Regular grooming of the breeding bull be practiced. In buffalo bulls regular shaving may be practiced.

Care and Management of Bullock

Bullocks are normally used for agricultural operations and or transport purpose. Some bullocks are ferocidus and so control them properly with nose rope or nose rings. The hooves of the bullocks should be provided with metal shoes to protect the hooves from wear and tear. The working hours for bullocks are recommended as follows:

a) Normal Work-6 hours of carting or 4 hours of ploughing.

b) Heavy Work-8 hours of carting or 6 hours of ploughing.

Sufficient roughages and 1-2 kgs of concentrates may be provided for feeding of bullocks during break period in works, the animal may be left for free grazing.

The bullocks are housed in separate sheds with sufficient space and protection from hot and cool conditions. Free access to drinking water is essential. Regular grooming of animals should be practiced.

Management of the Dairy Cow

There are thousands of dairy farms around the world with different farm plans and management strategies. Climate, market conditions, traditions, breeds etc. affect the running and planning of the business. If we try to categorise and generalise the world wide dairy operations, we relatively easily find three main "types". The American type (North and South America) is characterised by large loose-housing operations, total mixed ration feeding (TMR) and relatively many employees. However, dairy farms in Northeast US and parts of Canada differ from the typical American operation. There you find many smaller family farms with either loose-housing or stanchion barns. These operations are quite similar to the European type, which is characterised by relatively small operations where each cow is fed and treated individually.

The third type is mainly found in New Zealand and Australia. The dairy operations in this area are usually run in a very extensive way, relying to a great extent on grazing. Due to the climate and local restrictions, New Zealand and Australian dairy farmers do not rely on concentrate use so much. On top of that, the milk market is deregulated and the prices follow the world market price (WMP).

When you look closely at all these farm types, they share many underlying principles. Regardless of management strategy, farm size or local climate, the cow has to calf, produce milk, eat, be kept in good health status etc. The goal for any dairy farm should be to maximise profit per unit of milk within farm constraints and, in some parts of the world, within milk quotas. To be able to do so, we need to know how to manage the cow and how different production aspects interact with each other.

Feeding

As the dairy cows genetic potential is increasing, feeds and feeding strategies are becoming more and more important. It is well known that the amount of milk to be produced is highly influenced by the amount and quality of the feed given to the cow.

It is also possible to influence the milk composition through the feeding. As the cow normally experiences a shortage of nutrients in early lactation, it is of importance to feed the cow a well balanced diet in order to maximise the dry matter intake. An unbalanced diet increases the risk for metabolic disturbances and weight loss, which have a negative effect on the milk yield. Healthy cows will also make the transition from dry to peak easier.

The cow's largest compartment is the rumen. Together with the reticulum, it has a total volume of approx. 150.200 litres. In these two compartments, there are billions of micro-organisms. They help the cow to digest and utilise the nutrients in the feed. To reach a good feed utilisation, and in the end a high milk yield, the micro-organisms have to have optimal conditions.

Managing Feeding

One of the primary keys to a successful dairy operation is a good nutrition program. Not only is nutrition one of the highest input costs (about 50% of the total costs), but it also controls the results of milk production, reproduction and health. The basis in all feeding programs is the roughage and a common recommendation is that this shall account for 40.60 percent of the total dry matter intake.

A cows intake per day is limited, and it is therefore important to know the dry matter and nutritional content of both the roughage and concentrate. However, the cows daily dry matter intake is dependent on her stage of lactation. The intake will gradually increase after calving, to reach its peak approx. 6.12 weeks after calving.

When all the feedstuffs (including roughage) have been analysed, we have to make sure that each cow gets what she need in terms of energy, protein, minerals, vitamins and water. To do so there are a number of feeding strategies and systems to choose between. Their common trait is that the roughage intake is controlled through the distribution of concentrates. The more concentrates that are supplied, either separately or in a mix with the roughage, the less roughage the cow will eat.

Feeding Strategies

Ad Libitum Feeding

Ad libitum feeding simply means that the animals have free access to the feed and are allowed to eat as much as they want. The roughage is often fed ad libitum regardless of the strategy used for feeding the concentrates (flat rate feeding, challenge feeding and feeding to yield). However, by mixing roughage and concentrates and feeding it ad libitum, the total feed intake for a group of cows can be regulated by means of the degree of concentration of the feed. This strategy is common when using a mixer wagon. A disadvantage with ad libitum feeding is that it is not possible to control each cows individual feed intake, which increases the risk of both over-and underfeeding.

Flat Rate Feeding

Flat rate feeding is a strategy in which all cows are fed the same quantity of concentrates during the entire (or part of) lactation. The concentrate is restricted on a certain level, while the roughage is fed either ad libitum or in a mix. As described earlier, the cows energy and nutrient demand varies depending on the stage of lactation. Because of the fixed concentrate ratio, flat rate relies on fat mobilisation. The disadvantages with this strategy are the risk for metabolic disturbances (Ketosis) and the difficulty to reach high peak yields. However, the technique is simple and therefore the investment cost can be relatively low.

Challenge Feeding

While flat rate feeding relies on fat mobilisation, challenge feeding and feeding to yield aims to supply the cow with nutrients that are needed for the actual lactation stage.

Challenge feeding of concentrates takes place in the early lactation, when there is a risk of underfeeding. Every cow is given the maximum ration of concentrates that she can consume, without reducing her roughage intake below approx. 40.60 % of total dry matter intake. This is continued until she reaches peak, four to ten weeks after calving. By using this strategy, each individual cow is given the possibility to show her production potential.

Feeding to Yield

When the peak production is reached and the production starts to decline, there is a risk of overfeeding. By feeding the cow to yield, which means that each cow is fed concentrates individually according to her actual milk yield, the cows body condition is maintained and the feed is used as efficiently as possible. A successful implementation of this strategy means increased milk production with the feed cost well under control.

Feeding Strategy

Ad libitum:

- High risk of over/ underfeeding
- Waste of feed
- Difficult to reach high peak production

Flat rate feeding:

- Simple ration
- Risk of overweight cows
- Risk of underfed cows which may cause Ketosis
- Difficult to reach high peak production.

Challenge Feeding

Feeding to yield:

- Potential to reach a high production
- The cow is fed due to its demand
- Reduced risk for overweight cows
- Risk of pH-variation if the feeding is not spread over the whole day.

Feeding Systems

The layout of the feeding system and the equipment involved differs depending on the type of barn, stanchion or loose-housing, and the management strategy used on the farm. However, the principles of the feeding systems are generally the same and can be divided into the following four categories: Manual Concentrate Feeding, Computerised Concentrate Feeding, Total Mixed Ration (TMR) and Partly Mixed Ration (PMR).

Manual Concentrate Feeding

This system is common in stanchion barns but it is also found in loose-housing barns using manually operated in-parlour feeding. In a stanchion barn, the system makes individual feeding of both roughage and concentrate possible. However, in both loose-housing and stanchion barns, the roughage is usually fed ad libitum and the concentrate is fed individually.

Suitable feeding strategies depend on the barn layout. In a stanchion barn it is possible, with a varying grade of precision, to implement challenge feeding and feeding to yield strategies. A loose-housing barn with manual in-parlour feeding is more or less only suitable for flat rate feeding.

A drawback with manual systems is the risk of low accuracy. Practical experience has shown that there often are great differences between the intended ration and actual ration. The diagram below shows that, for example, if an animal should have 4.5 kg, the actual ration could range from 3 kg to 8 kg. This may not only result in costly feed waste and overweight cows, but will also have a negative effect on the milk production. A manual system is also labour demanding and requires more time to implement new feed rations, compared to a computerised system. All this means high feed and labour costs and lost milk revenue opportunities.

Computerised Concentrate Feeding

Instead of feeding the concentrates manually, they can be dispensed and controlled by a computer. In stanchion barns, a rail suspended feed wagon is normally used.

For loose-housing systems, there are two systems to choose from: in-parlour feeding or out-of-parlour feeding (or a combination of both).

The in-parlour feeding is simply a dispenser that can be operated by manual or automatic identification. Out-of-parlour feeders requires automatic identification (transponders). Farms with a long grazing period sometimes prefer to feed all concentrates in the parlour. A disadvantage with this is that large rations cause long eating times and are unhealthy for the rumen. An alternative to this is to use out-of-parlour feeders all year round and keep the cows housed for a couple of hours before letting them out on pasture again.

When out-of-parlour feeding or a rail suspended feed wagon (stanchion barns) is used, challenge feeding and feeding to yield can successfully be used as the feeding strategies. Each cows individual ration is programmed in a computer and when the cow enters the feed station, she is identified and the correct amount of feed is dispensed. The computer decreases the demand for labour and facilitates the implementation of new feed rations and the record keeping for each cow.

Total Mixed Ration (TMR)

In the Total Mixed Ration feeding system, the concentrates and roughage are mixed in a mixer wagon and usually fed ad libitum. The mix is often dispensed to the cow directly from the mixer wagon, but can also be distributed by band conveyors or with rail suspended feed wagons. The TMR system is most common in large loose-housing herds, but can also be used in stanchion barns.

There are mainly two ways to feed the cows with TMR: either feed the same ration to the whole herd (no grouping) or different rations to different groups. In non-grouping systems, only one mix is made and dispensed. The mix is usually composed to suit the high yielding cows, which makes the balance of the ration very important to avoid overweight cows. A way to improve the precision of the feeding is to use a grouping system. The cows can be grouped according to many criteria: yield, stage of lactation, first calf heifers etc. The drawback with grouping is that much of the rationality of the system is lost compared to the non-grouping system.

Much time must be spent moving the cows from one group to another as the production changes. Group changes also result in milk loss due to social adjustment. In addition, several mixes have to be prepared and dispensed to suit each group. However, if the herd is large enough, it should be possible to run an efficient and convenient TMR system with grouping. Disadvantages with all TMR systems are the relatively high amounts of feed waste and the fact that it is not possible to monitor the individual consumption of concentrates. The cost of feeding and storage equipment is also relatively high.

The main advantage is good rumen health, which results in few metabolic disturbances and possible effects on the milk production.

To obtain a well working TMR system and a high dry-matter intake, the ration requires much attention.

Partly Mixed Ration (PMR)

Partly Mixed Ration (PMR) is a feeding regime that combines TMR and individual feeding of concentrates. A mixer wagon is used to mix roughage and some of the concentrates. The concentrate level in the mix is adjusted to fit the lower yielding cows.

The high yielding cows are then fed extra concentrates by out-of-parlour feeders, in-parlour feeders or from a rail suspended feed wagon. The PMR system gives you the possibility to combine the advantages of a mixed ration with computerised feeding.

Milking

To illustrate an individual cows milk production, we normally plot the yields against time, which gives us the lactation curve illustrated in the picture below. As the diagram shows, the milk yield will rise during the first months after calving, which then is followed by a long period of continuous decline. The shape of the lactation curve will differ from individual to individual and from breed to breed. Feeding and management will also influence the shape and have a significant impact on the total amount of milk produced. Lactation is ideally 305 days, but in practice it is usually more, followed by a two-month dry period prior to the next calving. A cows milk yield is influenced by many factors, which are described more in detail in the DeLaval booklet Efficient Milking.

Peak Yield

Peak yield is the point where the cow reaches the highest milk production level during the entire lactation. Heifers peak at 70.75% of mature cows and second lactation cows peak at 90% of mature cows. Normally the peak is reached four to ten weeks after calving.

The time it takes to reach peak yield varies with many factors, for example breed, nutrition and yield potential. Higher producing animals tend to peak later than low producing ones. A high peak yield normally means a higher total yield. Research shows that each one kilogram increase in peak yield usually means an additional 100.200 kg of milk produced during the actual lactation. Reaching high peak yields requires a very well managed and balanced feeding programme.

Persistency

After the peak, milk production starts to decline by approximately 7.10% per month. The rate of decline is normally measured as the cows persistency. If a cows milk production falls by 7% per month after peak, it is more persistent than a cow that loses 10% per month. A general rule of thumb is that higher peak production leads to lower persistency. As for peak yield, the persistency is dependent on feeding and is therefore, to some extent, possible to influence. Persistency differs between cows, but a first lactating cow is normally more persistent than a second or third lactating cow.

Total and Daily Milk Yield

Close monitoring and evaluation of each cow or group are important ingredients when trying to reach a high level of total milk production (high peak and long persistency). However, it is important to remember that a dairy cow is not a stable milk producer. The milk yield differs from day to day and the relative variation can be as high as 6.8 % from one day to another. Cows milked three times or more a day usually have a lower variation than cows milked twice a day.

Managing Milk Yield Recording

The total milk yield is a good indicator of gross income from milk sales, which has a direct effect on the dairy farms income. With milk yield recording, the dairy farmer can monitor his day to day output and use it to control and monitor the production. Milk yields are also the most interesting variable when calculating the feed rations. Common methods for measuring milk yield are: bulk tank monitoring, monthly milk yield recording, recorder jars/stand alone meters and electronic milk yield registration. The milk yield records can then be used for:

- Calculating and follow-up of the nutrient intake
- Monitoring long-term production changes
- Indicating overall health disturbances.

Bulk Tank Monitoring

Bulk tank monitoring does not enable you to monitor the cows individually, but it can be used as an indicator of the overall herd management. To get the correct total yield, you have to account for the number of fresh, treated and dry cows plus the milk used for own consumption and calves.

Monthly Milk Yield Recording

Monthly recording is usually carried out by a milk yield recording organisation and was originally introduced for breeding purposes. Due to the fact that the recording normally takes place once a month, and each cows milk production varies day to day, it is not recommended to use this information to manage feeding. When receiving the data, it will be historical and not reflect the current situation.

However, the information can be used for trend analysis and give an idea of the peak and persistency for individual cows, which can be used for culling, breeding and general feeding decisions. As for bulk tank monitoring, the monthly milk yield recording does not give the dairy farmer an early indication of health problems.

Daily Milk Yield Recording

Daily milk yield recording can be considered as one of the most important decision aids for fine tuning the high producing herd. The milk yield can be measured at every milking by using either recorder jars or milk meters. If using recorder jars, all data has to be recorded manually (e.g. on paper or typed into a computer).

With milk meters connected to a processor or computer, you will have an automatic collection of data and a memory capacity that is much more accurate and efficient than the milkers. The processor contains a database where all individual milk yields and other relevant data are stored on a daily basis. This data provides the manager with exact and timely information of a cows or groups production on a day to day basis, which can be used to:

- Calculate and evaluate feed rations for individual cows or groups to maximise the milk production and avoid over and/ or under feeding.
- Identify cows in heat. Research shows that decline in morning milk might indicate oestrus.
- Identify cows with potential health problems early. For example, ketosis is related to a gradual decrease in milk production before it can be diagnosed.
- Calculate the actual lactation curve instead of estimating.
- Group cows by yield, which results in faster milking.
- Evaluate long-term milk production for individual cows to plan breeding and culling.

Reproduction

Reproduction is a necessary and important part of milk production. Without regular calvings in the long run, there will be difficulties in producing the desired amount of milk. It is also important to produce sufficient heifer calves as replacement animals and allow herd size to be maintained or expanded.

Today it is most common to use artificial insemination (AI) instead of a bull. AI increases the control over the breeding and enables the dairy farmer to use sperm from all over the world, which increases the genetic gain. However, at the same time, it increases the need for good planning and well structured working routines.

Another way to get a cow pregnant is to use the embryo transfer technique. So far this is relatively uncommon, but it is expected to increase in importance. The main advantage with embryo transfer is that it is possible to generate more calves from a good cow than it is with AI.

The Oestrus Cycle

As long as a cow or heifer is not pregnant she will normally have a 21-day oestrus cycle. The length of the oestrus cycle may vary but it usually ranges from about 17 to 24 days. A heifers oestrus cycle is normally slightly shorter than a cows. The cycle will continue until the cow is pregnant. After calving, cows normally undergo a 20 to 30-day period when oestrus cycles do not occur.

The oestrus cycle is controlled by a complex system involving different hormones produced in the brain and ovary. The picture below shows a simplified picture of how two of these hormones, oestrogen and progesterone vary depending on where in the cycle the cow is.

Some cows do not follow the normal oestrus cycle. For example, a cow can be unoestrus, which means her ovaries do not function with the regular 17 to 24-day cycle and are therefore not observed in heat. Other cows may suffer from ovarian cysts. These cows will show heat at very short intervals and the period that they are in heat will last three to four days.

Reproductive Management

Research has shown that suboptimal calving interval causes large economic losses to dairy farmers, second in importance only to mastitis. Milking cows in late lactation are less profitable due to the decline in production. A long calving interval means milking less profitable cows due to the decline in production, fewer calves and too many cows with low feed conversion efficiencies. Therefore, successful reproductive management has a significant impact on the herds overall performance and the net income.

Well managed reproduction also reduces the risk of expensive involuntary culling. Information from UK and USA shows that the cost for a one day extended calving interval ranges from £1.5 to £3 per cow.

The time of the calving interval that is possible to influence by management is the open days, which are determined by the voluntary waiting period (VWP) and the breeding window (BW). A very common reason for undesired long calving intervals are missed heats. With improved heat detection rate (HDR) and conception rate (CR), through better management and improved timing of in seminations, it is possible to obtain a significantly shorter calving interval.

Detecting Heat

The most sexually intensive period of the oestrus cycle is during standing heat, which lasts for approximately 18 hours. In loose-housing herds, this period is indicated by the cow in heat standing immobile when mounted by another cow or bull. Other signs of heat are:

- Bellowing
- Increased activity
- Walking the fence line
- Licking/Sniffing
- Swelling and reddening of the vulva
- Mounting other cows
- Lower milk yield
- Reduced feed intake.

The duration of heat varies from animal to animal, but approximately 10 to 12 hours after the end of standing heat, the egg is released (ovulation) and the heat ends.

Manual Heat Detection

Manual heat detection relies on manual observations in the barn. The cows and heifers shall be observed for oestrus two to three times per day and all observed heats be recorded whether the animal is bred or not. The records provide the manager with information to anticipate future heats, which will make it easier to distinguish if a cow is in heat or not. Most mounting occurs between 6 p.m. and 6 a.m., and it is therefore worth trying to check for heat during these hours. To facilitate the planning and record keeping, a cow calendar is often used. This can be either manual or computerised.

The disadvantage with manual detection is that it is very time demanding and requires people with the ability to observe the right signs. This is particularly important when there are no distinct signs of heat. By using progesterone testing or other tools on the market, the detection rate can be improved.

Automatic Heat Detection

Another way to identify cows in heat is to monitor their activity. During heat the activity can increase up to 8 times compared to the normal level.

The activity can automatically be recorded by using activity meters attached to the neck or leg.

By comparing the activity with the last observed heat, actual milk yield and feed consumption, a reliable indicator of heat is obtained. The automation will generate a significant time saving and improved calving interval through better heat detection.

Timing of Insemination

With artificial insemination, the timing of the insemination becomes important. The optimum time for insemination depends on when the ovulation occurs in relation to the heat and for how long the sperm is viable. Most sperm remain viable for about 24 hours. The ovums life is only about four hours and is the most critical. It is therefore preferable that viable sperm is present in the salpin during ovulation. There are mainly two rules for the timing of insemination. Traditionally the a.m.-p.m. rule was the one dairy producers followed. This rule dictates that cows and heifers first observed in heat in the morning should be bred late in the afternoon. Likewise, cows and heifers first observed in heat in the afternoon should be bred the following morning. This is still a good rule, but many producers have now successfully gone to once-a-day insemination. This rule dictates that cows and heifers first observed in oestrus in the afternoon or the following morning, should be bred late that morning.

Herd Health

Disturbances and diseases will always occur in a herd, but in order to limit the economic loss, it is important to keep them under control. Good health and health management have a significant effect on a dairy farms net result. As the number of different disturbances is great, only the most common ones will be covered here.

Common Diseases and Disturbances

Retained placenta: This is a condition where the foetal membranes are not completely expelled. Retained placenta increases the risk for a whole host of other health problems (e.g., uterus infection, ketosis, and displaced abomasum). Usually a retained placenta is associated with difficult calving, a mineral imbalance (e.g., potassium, calcium) prior to calving, or over conditioning.

Milk Fever: Milk fever occurs when there is a shortage of calcium in the blood. This also usually results from a mineral imbalance (e.g., potassium, calcium) prior to calving. The effect of milk fever normally starts with reduced feed intake, followed by difficulties for the cow to move. Eventually the cow becomes paralysed and, if not treated, may die within a couple of hours. However, treatment with calcium usually has a good effect and in most cases, the cow will survive.

Ketosis: Ketosis occurs when cows begin to milk but have a shortage of energy. These cows will use fat to support milk production, but their liver cannot convert the fat to energy fast enough. Ketone bodies accumulate in the blood causing ketosis. Treatments generally consist of rapidly digested sources of glucose such as propylene glycol or molasses.

Left-Side Displacement of Abomasum (LDA): This may occur in relation to calving. The fourth stomach (abomasum) migrates from the right side of the cow to the left side. It is thought that this occurs with low fibre levels in the diet, physical stress such as slippery floors, and secondarily to other problems such as ketosis.

Acidosis: There are a few health problems directly associated with feeding, but high levels of concentrates might cause acidosis. The rumen pH drops below 6.0 for a long period of time causing numerous problems. One result of acidosis is hoof problems. Cows that have acidosis may experience hoof ulcers and a form of founder.

Mastitis: Mastitis is the most common and costly disease in dairy herds. It is an inflammation in the mammary gland which can be caused by bacterial infections or trauma. Cows with clinical mastitis are relatively easy to detect for the farmer. The symptoms are clotting and discoloration of the milk, and the gland becomes hard, red or swollen and in severe cases the cow has fever and loss of appetite. Cows with subclinical mastitis will have no outward signs of mastitis but their mammary gland will still be infected and the somatic cell count will be high.

Lameness: Hoof problems or lameness is characterised by damage to sensitive laminae of the hooves and it is reported to be the third most common cause of culling behind reproduction and mastitis. Hard and slippery floors, decreased exercise and metabolic disturbances (especially acidosis) negatively affect the occurrence of hoof problems. Estimates are that cattle that become lame and are not tended to, can experience a 20% loss in milk production.

Managing Herd Health

The economic losses due to health disturbances can be attributed to one or more of the following factors:

- Less efficient production and higher veterinary costs
- Reduced slaughter value and idle production factors
- Lost future income.

Lost future income occurs when the cows have to be replaced (culled) before reaching their economically optimal age. This varies from individual to individual, but it is basically the time when it is more profitable to replace a cow than to keep her. Culling is usually the eventual outcome of health problems and culling rates are normally found to be between 25.35%. Cows that die on the farm are part of this culling rate and should represent less than 3% of the total herd.

The reduction of losses attributed to diseases and the promotion of positive welfare and health should be one of the main goals in dairy herd management. A helpful tool for maintaining healthy cows is body condition scoring (BCS). Other important aspects that might affect health, if not looked after and maintained, are barn environment, milking routines and milking equipment.

Body Condition Scoring

Body condition scoring (BCS) can be used to troubleshoot problems and improve the health and productivity of the dairy herd. Overweight cows (overconditioned) are more susceptible to metabolic problems, infections and noninfectious health problems. Research shows that overweight cows are more likely to be affected by, for example, mastitis, retained placenta, ketosis and lameness. They are also more likely to have difficulties at calving. Thinness (underconditioning) can lower the milk production and fat content because of insufficient energy and protein reserves.

The body condition scoring system uses a one to five, scale with one representing very thin cows and five very overweight cows. An ideal cow has a body condition score of about 3.5, but the system is designed to have cows at certain stages of lactation at certain body conditions. Important parts to look at when scoring are: chine, loin, rump, hip, tail head and pin bones.

Score	*Condition*
1	Skin and bones
2-2.5	Severe negative energy balance on cow in early lactation. Risk of production loss.
2.5-3	High producer in early lactation.
3-3.5	Milking cow in good nutrient balance.
3.5-4	Late lactation and dry cow in good condition.
4	Overconditioned. Potential calving problems if dry.
5	Severely overconditioned. Risk of fat cow syndrome.

Dry cows should neither gain nor lose weight and their body condition score should be around 3.5. If they have too much fat (>4.0) they tend to have many problems after calving. The reason is that all of this body fat must move through the liver before it can be used for milk production. This is not bad if they use the fat slowly, but if they begin to lose weight quickly, fat will accumulate in the liver and a condition known as fatty liver develops. These cows usually perform poorly and are more prone to die. After calving, cows should lose less than one point before they begin to gain weight again. Cows that lose more than one point tend to have more reproductive problems.

From peak milk production to dry off, cows should gain back the body condition that was lost before the peak was reached. It is common for cows that have problems (e.g., ketosis, hoof problems, mastitis etc.) to lose one body condition score within the first two weeks of lactation. Ideally, the body condition score of a cow should be taken whenever she is handled. At least body condition scores should be taken at freshening, breeding and dry off. If over-or underconditioning occurs, the feeding strategy for an individual cow or group has to be evaluated and feed rations may have to be recalculated and corrected to satisfy the cow's demand.

Preventing and Controlling Mastitis

Mastitis needs extra attention, as it is the most common disease among dairy herds. It is not unusual that up to 40% of the herd is infected by clinical (with visible symptoms) or subclinical (often without symptoms) mastitis. All these cases are associated with many costs.

With good management routines, there are great opportunities to reduce the number of cases. This has a direct effect on production levels and costs. For example, if a farm is able to reduce the somatic cell count from 200 000 to 100 000, it means an increased average production of 0.7 kg per cow and day. In a 100 cow herd, this gives an additional milk production of 0.7 X 365 X 100 = 25 550 kg of milk per year.

The management must focus on actions that prevent new infections and reduces the duration of each infection. This includes proper maintenance of bedding and stalls, hygiene procedures during milking and frequent checking of the milking equipment. Dairy cows require a comfortable, clean, dry and draft-free environment. Comfort can include everything from feed space and walking surfaces to air and resting areas. However, even if the physical facilities are well planned, the daily work routines are of great importance to prevent as many health problems as possible. This includes not only mastitis, but also hoof problems (lameness). The way to avoid many of these cases is to regularly trim the hoofs and make sure that the surfaces in the barn are dry and free from mud.

Important Routines for Preventing Mastitis

With good milking routines and adequate milking equipment, the number of new mastitis cases will be significantly lowered. But even so, there will still be cases that need treatment. To avoid excessive costs associated with these cases, the duration of each case has to be reduced through:

- Prompt treatment of clinical cases; early detection, appropriate therapy and recording of clinical cases.
- Culling cows with chronic mastitis.
- Regularly checking the somatic cell count on all cows (High SCC indicates subclinical mastitis).

- Culling persistent offenders, e.g. cows having three cases in one quarter or five cases in mixed quarters in one lactation.

Milk Quality

Milk consumers require healthy and safe milk that is free from contamination, unpleasant smell etc. Processing also demands for milk to meet certain standards. For example, cheese making is very sensitive to spores and antibiotics. The requirements on the quality has increased over the time and will continue to do so. To meet this, more and more dairies put pressure on the milk producers. This is normally reflected in the payment scheme, where producers that deliver milk that does not meet the required standards have to pay penalties. There are also examples of processors paying bonuses to producers that meet the highest standards. As this makes the quality of the milk a determinant of the producers final price, it has to be considered in the daily routines.

The dairies use different parameters to check the quality of the milk. The exact level for each parameter varies from company to company. Therefore it is worth taking a closer look at the demands and standards you have to meet. Examples of tests carried out on delivered milk are:

- Fat and protein content
- Bacteria count (Bactoscan)
- Smell and taste
- Residues from antibiotics and other Medicines
- Freezing point (detects water content)
- Somatic Cell Count
- Spores (especially important for cheese making).

Managing Milk Quality

There are many factors that affect milk quality and some are easier to control than others. With well structured daily routines and an awareness among the personnel, it is possible to control and improve the situation considerably. Below you find a list of actions that all will have a positive effect on the overall quality and the final price of the milk:

- High quality water in sufficient amounts
- Feeds with high hygienic quality

- A well balanced ration with suitable ingredients
- Good overall hygiene
- A good and well maintained barn and parlour environment
- Prompt treatment of diseases
- Do not deliver colostrum milk
- Avoid delivering milk from cows with high cell count
- Do not deliver milk containing residues of antibiotics, sulpha etc.
- Do not add any water to the milk
- Make sure no detergent and/or other chemical residues get into the milk
- Ensure good ventilation to avoid undesired smell
- Make sure the milking equipment is cleaned properly
- Cool the milk directly after milking (Avoid freezing!).

3

Chemical Constituents of Milk

Composition of Milk

Milk from different animals varies somewhat in the proportion of the different constituents. All milk contains a high proportion of water, cow's milk averaging about 87 per cent. The milk from different breeds of cattle varies considerably in composition. The milk of individual cows of the same breed also varies in the proportion of the different constituents. This may be partly due to environment, partly to inheritance, and partly to individuality. Milk may vary particularly in fat content from one milking to the next. The percentage of fat increases during the milking period; that is, the first milk obtained is not so rich in fat as the last portions of milk.

The fat varies for different breeds from about 3.5 per cent for Holstein to about 5 per cent for Guernsey and Jersey.

The protein varies from about 3.3 per cent for Holstein to 4.0 per cent for Guernsey and Jersey. It parallels the fat content somewhat, being highest in the breeds having a high fat content and lowest in those having a low fat content.

The lactose does not vary so much with the different breeds as the fat and protein. The lactose content is from about 4.65 to 5 per cent.

The ash varies from about 0.68 to 0.75 for the different breeds.

The constituents of milk that are most important in food preparation are enzymes, vitamins, pigments, salts, sugar, fat, and proteins. Enzymes. The enzymes of cow's milk are reported as follows by Rogers; proteinases, lactase, diastase, lipase, salolase, catalase, peroxidase, and aldehydrase. Rogers states that the proteolytic enzyme, galactase, brings about slow decomposition of milk proteins into peptones, amino acids, and ammonia. Vitamins. All the vitamins recognized at the present time are contained in milk. Some are present in comparatively large and others in smaller amounts.

Pigments. The appearance of milk is white. This is due to light rays reflected by the colloidally dispersed constituents of the milk, the calcium caseinate, and calcium phosphate.

Milk contains two classes of yellow or orange pigments. The water-soluble pigment, which imparts a yellow colour with a green fluorescence to the whey of milk, was formerly called lactochrome. A name recently suggested for this pigment is lactoflavin. It is regarded as one flavin of a specific group, collectively to be called lyochromes. It is possible that lacto-flavin is composed of more than one pigment. Rogers says "lactoflavin forms compounds with saccharides, proteins, and purines (uric acid). These compounds possibly either occur naturally in milk or readily form during isolations, thus accounting for the several lactoflavins isolated from milk." It is probable that the pigment lactoflavin is one of the five fractions of vitamin G (B2). Milk is relatively rich in this vitamin.

A fat-soluble pigment, carotene, found in the fat gives the milk a more or less yellow tinge, which is more pronounced as the fat particles become more concentrated and form cream The group of pigments called caroti-noids, which include carotene, xanthophyll, and related pigments, has been described in the chapter on fruits and vegetables. The chief pigment of butter fat is the carotene, but little xanthophyll being found. The depth of colour depends upon the amount of pigment present.

The colour of carotene in solution varies from yellow to orange and to a deep red-orange as the concentration increases. The amount of carotene found in the butter fat depends upon the extent of carotene in the food of the cow. Green grasses, hay cured to retain its green colour, green corn, and carrots are rich in carotene. The carotene content of milk fat is less rich during the winter months, if the food of the cow is poor in carotene during this period. Only in cow's milk is the fat extensively pigmented. With the exception of the fat of human milk, which is often pigmented, the fat of the milk of other animals is either devoid of or contains little pigment.

Salts. Milk contains salts of potassium, sodium, magnesium, calcium, phosphates, chlorides, and citrates. Traces of sulfates and carbonates are found. Iron is present in small amount. Iodides are also found in small amounts, the amount being greater in some localities than in others. Iodides may be easily transmitted from the feed to the milk. Supplee and Bellis have found copper to average about 0.52 part per million in freshly drawn milk.

Brickner has reported that milk contains 3.6 to 5.6 parts of zinc per million parts of milk. Manganese in normal milk averages 0.02 to 0.06 parts per million. The greater part of the sulfur is found in the milk proteins. Barger and Coyne state that part of the sulfur in milk is found in the amino acids methionine and cystine of the proteins, but that all of the sulfur is not accounted for.

The salts of milk are found in milk in solution, in the colloidal state, and in combination with the proteins. The exact chemical combinations of the different salts in the milk are not fully determined. For this reason different authorities report different salt combinations. Thus formulas imply definite combination, whereas there is a complex salt equilibrium in milk, which has not been satisfactorily worked out. The citrates, the combinations of which may be trisodium and tripotassium citrate, tricalcium and trimagnesium citrate, are probably entirely in solution. The possible chlorides of potassium, sodium, and calcium are also in solution. Some authorities believe that the phosphates are present chiefly as dicalcium phosphate, CaHPO4 others believe that tricalcium phosphate, Ca3(PO4)2, is the principal phosphate combination. The phosphates are partly in solution, with the greater portion in colloidal dispersion. When the particles of dicalcium phosphate are heated they become aggregated and partially precipitated. Calcium and magnesium are in combination with the casein to form calcium and magnesium caseinates. Zoller states that there may be traces of sodium and potassium caseinates.

Lactose. The solubility of lactose and its properties may be found in the chapter on sugar. It is caramelized by heat at rather low temperatures.

Fat. Butter fat is composed of glycerol and fatty acids. Fatty acids of both the saturated and unsaturated series are present. The relative percentage of the unsaturated fatty acids varies with the feed, averaging higher in summer than in winter. Dean and Hilditch state that the oleic and linoleic acids increase by 4 per cent (mols), with a parallel diminution in butyric and stearic acids when cows return to pasture. They also report that with increased age of the cow the unsaturated acids increase at the expense of palmitic acid, which was lowered from

29 to 22-23 per cent by weight of the total fat in the four years of observations made on milk from the same group of cows. The saturated fatty acids are butyric, caproic, caprylic, capric, lauric, myristic, plamitic, and stearic.

Arachadonic acid has been reported absent by some investigators. Butter fat contains a higher proportion of the first-named saturated acids than other food fats. The first ones in the series are quite volatile with steam, the volatility decreasing with increase in molecular weight of the acid. Hence, when butter is heated for several minutes, the percentage of the lower saturated fatty acids may be decreased. The unsaturated fatty acids include oleic, linoleic, and arachidic.

Formerly it was thought that the particular flavour of butter was due to the greater proportion of the lower saturated fatty acids, but now it is known that butters with satisfactory flavour and aroma (Michaelian, Farmer, and Hammer) contain considerable quantities of acetylmethyl-carbinol plus diacetyl. The acetylmethylcarbinol in a pure condition is odorless, but by bacterial action it is changed to diacetyl, which in concentrations of 0.0002 to 0.0004 per cent in or added to neutral butter gives a characteristic aroma.

Size of Fat Globules

The fat in milk is found in small globules. These fat globules which are microscopic in size are suspended in the milk. They vary in size from 0.10μ to 22.0μ. Most of the globules are less than 10μ and average about 3μ in diameter. The size of the fat globules varies (1) with different breeds, being larger in milk from Jersey and Guernsey than in milk from other breeds, (2) with the lactation period, decreasing in size with length of the lactation period, and (3) with feed. Dry feed tends to decrease the size of the globules, succulent feed to increase their size.

Creaming

The specific gravity of the fat globules is less than that of the fluid of the milk. Hence there is a tendency for them to rise to the surface to form a cream layer. The extent of creaming depends upon several factors, such as the size of the fat globules, temperature of the milk, acidity, physical state of the fat, etc. Creaming occurs more rapidly in milk when the fat globules are quite large than when they are smaller. In rising to the top of the milk the globules of fat clump together, and this clumping increases the tendency for them to rise to the surface of the milk. As the larger clumps rise they may carry many of the smaller globules to the surface. Therefore, clumping not only aids the completeness of creaming but also the rate, for the rate is more rapid when the fat particles clump quickly. Rogers states that the factors affecting clumping are "temperature, the acidity, the fat content and its degree of dispersion, the degree of agitation, and the fluidity of the system.

The fat content, the degree of agitation and the fluidity of the system determine the probability of collisions of the globules." The tendency of the fat globules to clump is greatest when the milk is cooled rapidly to 7° to 8°C, but if the fat globules become solid at the low temperature before creaming is allowed to take place the rate of creaming is retarded. The temperature that favours clumping is also best for whipping the cream. Most cream is separated from the milk by mechanical means. The fat particles left in the milk after separating the cream are those less than 1u and those between 1 and 2u in diameter.

Butter

The fat globules in cow's milk are suspended in the milk and thus do not form a permanent emulsion, though they may be so reduced in size by homogenization that they form a permanent emulsion. Of the different theories formulated for explaining the manner in which emulsions are stabilized the adsorption film theory is usually connected with milk. Substances that lower surface tension tend to collect at the interface between two non-miscible systems. Proteins tend to lower the surface tension, hence tend to collect at the interface.

The layer or film around the fat globules probably consists of adsorbed calcium caseinate, with some lactal-bumin, globulin, and calcium phosphate. This membrane may be weakened or broken in various ways. When milk is heated slowly, the membrane surrounding some of the fat globules may be broken and a number of globules may coalesce. Sometimes milk that has been heated and then cooled has a more oily appearance due to this coalescing of the fat globules. The membrane surrounding the fat particles may be broken by mechanical agitation. Formation of butter in churning is brought about in this manner.

There are Two Theories Regarding Butter Formation: One is that the emulsion is reversed from the type found in the milk and that the butter is a water-in-fat emulsion. The other view is that butter is formed by packing the fat globules into a compact mass, and that water and air are enmeshed during this process. Temperature and the formation of a foam are both important in churning. At a temperature above 65°C. there is no aggregation of fat particles. Below 65° the tendency to clump increases and is at a maximum at 7° to 8°C. A favourable temperature for butter formation is below 24° and above 10°C. At temperatures below 4°C. the fat globules do not adhere to each other and aggregation does not take place. At temperatures above the melting point of butter, no butter is formed.

The fat particles tend to clump at the liquid/air interface, so that air beaten in during the churning process accelerates clumping of the fat. Butter may be churned from sweet or sour cream. The flavour of butter from the sweet cream is milder and different from that of the sour-cream butter. Many housekeepers prefer the sweet-cream butter for table use and the sour-cream butter for cooking.

The adsorbed films surrounding the fat globules may also be destroyed by the addition of acid or alkali. It is by these methods that the fat is set free for a quantitative determination. In the Babcock test, acid is used for liberating the fat globules; in the Hoyberg test alkali is used. The Bab-cock, or some modification of it, is the one usually employed in estimating the fat content of milk and cream.

Protein

The chief proteins found in milk in order of their decreasing amounts are casein, lactalbumin, and lactoglobulin.

Casein

Casein belongs to the group of phosphoproteins. The form in which the phosphorus exists in the casein is not definitely known, but it is believed to be present in the form of combined phosphoric acid. Casein forms about 3 per cent of cow's milk.

At its isoelectric point, which is pH 4.6, casein is nearly insoluble in water. Casein is amphoteric and forms salts with acids and alkalies. Fresh milk has a reaction of about pH 6.6, so that the casein is present in the milk as salts of bases and is found as calcium and magnesium caseinates. All the alkali caseinates are soluble in water, though the salts of the alkaline earths are less soluble than the alkali ones. Loeb states that below pH 4.6 the casein chloride, casein acetate, and casein lactate are very soluble in water, but casein sulfate and casein oxalate are difficultly soluble.

According to Zoller, pure casein when heated in water begins to imbibe water at 80° to 90°C. and becomes plastic. In this form it can be molded and shaped. Upon cooling it becomes very hard.

Casein can be precipitated from milk by bringing the milk to the isoelectric point of casein. Coagulation of casein will be considered later.

Lactalbumin

The proportion of lactalbumin in milk is much lower than that of casein. It forms about 0.50 per cent of cow's milk. Its isoelectric point is pH 4.55. Since the reaction of fresh milk is about pH 6.6, it is on the alkaline side of the isoelectric point of lactalbumin. Thus is it possible that the lactalbumin is found combined as salts of bases, such as calcium and magnesium albuminates. Osborne and Wakeman think it is uncom-bined. Lactalbumin is soluble in water, and is coagulated by heating in solution to a temperature of about 70°C. Coagulation may not be complete at this temperature. Palmer states that the lactalbumin is more highly dispersed than the other colloidal constituents of the milk.

Lactoglobulin

Lactoglobulin occurs in milk in very small quantities, about 0.05 per cent of cow's milk. Lactoglobulin is insoluble in distilled water, but it is soluble in dilute solutions of strong bases or acids, and in dilute salt solutions. It is coagulable by heat.

Physical and Chemical Properties of Milk

The role of milk in nature is to nourish and provide immunological protection for the mammalian young. Milk has been a food source for humans since prehistoric times; from human, goat, buffalo, sheep, yak, to the focus of this section-domesticated cow milk (genus Bos). Milk and honey are the only articles of diet whose sole function in nature is food. It is not surprising, therefore, that the nutritional value of milk is high. Milk is also a very complex food with over 100,000 different molecular species found. There are many factors that can affect milk composition such as breed variations (see introduction, cow to cow variations, herd to herd variations-including management and feed considerations, seasonal variations, and geographic variations. With all this in mind, only an approximate composition of milk can be given:

- 87.3% water (range of 85.5%-88.7%)
- 3.9 % milkfat (range of 2.4%-5.5%)
- 8.8% solids-not-fat (range of 7.9-10.0%):
  - protein 3.25% (3/4 casein)
  - lactose 4.6%
  - minerals 0.65%-Ca, P, citrate, Mg, K, Na, Zn, Cl, Fe, Cu, sulfate, bicarbonate, many others
  - acids 0.18%-citrate, formate, acetate, lactate, oxalate
  - enzymes-peroxidase, catalase, phosphatase, lipase
  - gases-oxygen, nitrogen
  - vitamins-A, C, D, thiamine, riboflavin, others.

The following terms are used to describe milk fractions:

- Plasma = milk-fat (skim milk)
- Serum = plasma-casein micelles (whey)
- solids-not-fat (SNF) = proteins, lactose, minerals, acids, enzymes, vitamins
- Total Milk Solids = fat + SNF.

Not only is the composition important in determining the properties of milk, but the physical structure must also be examined. Due to its role in nature, milk is in a liquid form. This may seem curious if one takes into consideration the fact that milk has less water than most fruits and vegetables. Milk can be described as:

- an oil-in-water emulsion with the fat globules dispersed in the continuous serum phase
- a colloid suspension of casein micelles, globular proteins and lipoprotein partilcles
- a solution of lactose, soluble proteins, minerals, vitamins other components.

Looking at milk under a microscope, at low magnification (5X) a uniform but turbid liquid is observed.

At 500X magnification, spherical droplets of fat, known as fat globules, can be seen. At even higher magnification (50,000X), the casein micelles can be observed. The main structural components of milk, fat globules and casein micelles, will be examined in more detail later.

Milk Lipids-Chemical Properties

The fat content of milk is of economic importance because milk is sold on the basis of fat. Milk fatty acids originate either from microbial activity in the rumen, and transported to the secretory cells via the blood and lymph, or from synthesis in the secretory cells.

The main milk lipids are a class called triglycerides which are comprised of a glycerol backbone binding up to three different fatty acids. The fatty acids are composed of a hydrocarbon chain and a carboxyl group. The major fatty acids found in milk are:

Long Chain:

- C14-myristic 11%
- C16-palmitic 26%
- C18-stearic 10%
- C18:1-oleic 20%.

Short chain (11%):

- C4-butyric*
- C6-caproic
- C8-caprylic
- C10 – capric.

* butyric fatty acid is specific for milk fat of ruminant animals and is resposible for the rancid flavour when it is cleaved from glycerol by lipase action.

Saturated fatty acids (no double bonds), such as myristic, palmitic, and stearic make up two thirds of milk fatty acids. Oleic acid is the most abundant unsaturated fatty acid in milk with one double bond. While the cis form of geometric isomer is the most common found in nature, approximately 5% of all unsaturated bonds are in the trans position as a result of rumen hydrogenation.

Triglycerides account for 98.3% of milkfat. The distribution of fatty acids on the triglyceride chain, while there are hundreds of different combinations, is not random.

The fatty acid pattern is important when determining the physical properties of the lipids. In general, the SN1 position binds mostly longer carbon length fatty acids, and the SN3 position binds mostly shorter carbon length and unsaturated fatty acids. For example:

- C4-97% in SN3
- C6-84% in SN3
- C18-58% in SN1.

The small amounts of mono-, diglycerides, and free fatty acids in fresh milk may be a product of early lipolysis or simply incomplete synthesis. Other classes of lipids include phospholipids (0.8%) which are mainly associated with the fat globule membrane, and cholesterol (0.3%) which is mostly located in the fat globule core.

Milk Lipids-Physical Properties

The physical properties of milkfat can be summerized as follows:

- density at 20° C is 915 kg m(-3)*
- refractive index (589 nm) is 1.462 which decreases with increasing temperature
- solubility of water in fat is 0.14% (w/w) at 20° C and increases with increasing temperature
- thermal conductivity is about 0.17 J m(-1) s(-1) K(-1) at 20° C
- specific heat at 40° C is about 2.1kJ kg(-1) K(-1)
- electrical conductivity is <10(-12) ohm(-1) cm(-1)
- dielectric constant is about 3.1.

*the brackets around numbers denote superscript

At room temperature, the lipids are solid, therefore, are correctly referred to as "fat" as opposed to "oil" which is liquid at room temperature. The melting points of individual triglycerides ranges from-75° C for tributyric glycerol to 72° C for tristearin. However, the final melting point of milkfat is at 37° C because higher melting triglycerides dissolve in the liquid fat. This temperature is significant because 37° C is the body temperature of the cow and the milk would need to be liquid at this temperature. The melting curves of milkfat are complicated by the diverse lipid composition:

- trans unsaturation increases melting points
- odd-numbered and branched chains decrease melting points.

Crystallization of milkfat largely determines the physical stability of the fat globule and the consistency of high-fat dairy products, but crystal behaviour is also complicated by the wide range of different triglycerides. There are four forms that milkfat crystals can occur in;

alpha, ß, ß ' 1, and ß ' 2, however, the alpha form is the least stable and is rarely observed in slowly cooled fat.

Milkfat Structure-Fat Globules

More than 95% of the total milk lipid is in the form of a globule ranging in size from 0.1 to 15 um in diameter. These liquid fat droplets are covered by a thin membrane, 8 to 10 nm in thickness, whose properties are completely different from both milkfat and plasma. The native fat globule membrane (FGM) is comprised of apical plasma membrane of the secretory cell which continually envelopes the lipid droplets as they pass into the lumen.

The major components of the native FGM, therefore, is protein and phospholipids. The phospholipids are involved in the oxidation of milk. There may be some rearrangement of the membrane after release into the lumen as amphiphilic substances from the plasma adsorb onto the fat globule and parts of the membrane dissolve into either the globule core or the serum. The FGM decreases the lipid-serum interface to very low values, 1 to 2.5 mN/m, preventing the globules from immediate flocculation and coalescence, as well as protecting them from enzymatic action.

It is well known that if raw milk or cream is left to stand, it will separate. Stokes' Law predicts that fat globules will cream due to the differences in densities between the fat and plasma phases of milk. However, in cold raw milk, creaming takes place faster than is predicted from this fact alone. IgM, an immunoglobulin in milk, forms a complex with lipoproteins.

This complex, known as cryoglobulin precipitates onto the fat globules and causes flocculation. This is known as cold agglutination. As fat globules cluster, the speed of rising increases and sweeps up the smaller globules with them. The cream layer forms very rapidly, within 20 to 30 min., in cold milk.

Homogenization of milk prevents this creaming by decreasing the diameter and size distribution of the fat globules, causing the speed of rise to be similar for the majority of globules. As well, homogenization causes the formation of a recombined membrane which is much similar in density to the continuous phase.

Recombined membranes are very different than native FGM. Processing steps such as homogenization, decreases the average diameter of fat globule and significantly increases the surface area.

Some of the native FGM will remain adsorbed but there is no longer enough of it to cover all of the newly created surface area. Immediately after disruption of the fat globule, the surface tension raises to a high level of 15 mN/m and amphiphilic molecules in the plasma quickly adsorb to the lipid droplet to lower this value. The adsorbed layers consist mainly of serum proteins and casein micelles.

Fat Destabilization

While homogenization is the principal method for acheiving stabilization of the fat emulsion in milk, fat destabilization is necessary for structure formation inbutter, whipping cream and ice cream. Fat destabilization refers to the process of clustering and clumping (partial coalescence) of the fat globules which leads to the development of a continuous internal fat network or matrix structure in the product. Fat destabilization (sometimes "fat agglomeration") is a general term that describes the summation of several different phenomena. These include:

Coalescence: an irreversible increase in the size of fat globules and a loss of identity of the coalescing globules;

Flocculation: a reversible (with minor energy input) agglomeration/clustering of fat globules with no loss of identity of the globules in the floc; the fat globules that flocculate; they can be easily redispersed if they are held together by weak forces, or they might be harder to redisperse to they share part of their interfacial layers;

Partial coalescence: an irreversible agglomeration/clustering of fat globules, held together by a combination of fat crystals and liquid fat, and a retention of identity of individual globules as long as the crystal structure is maintained (i.e., temperature dependent, once the crystals melt, the cluster coalesces). They usually come together in a shear field, as in whipping, and it is envisioned that the crystals at the surface of the droplets are responsible for causing colliding globules to stick together, while the liquid fat partially flows between they and acts as the "cement". Partial coalescence dominates structure formation in whipped, aerated dairy emulsions, and it should be emphasized that crystals within the emulsion droplets are responsible for its occurrence.

Milk Lipids-Functional Properties

Like all fats, milkfat provides lubrication. They impart a creamy mouth feel as opposed to a dry texture. Butter flavour is unique and is derived from low levels of short chain fatty acids. If too many short chain fatty acids are hydrolyzed (separated) from the triglycerides, however, the product will taste rancid. Butter fat also acts as a reservoir for other flavours, especially in aged cheese. Fat globules produce a 'shortening' effect in cheese by keeping the protein matrix extended to give a soft texture. Fat substitutes are designed to mimic the globular property of milk fat. The spreadable range of butter fat is 16-24° C. Unfortunately butter is not spreadable at refrigeration temperatures. Milk fat provides energy (1g = 9 cal.), and nutrients (essential fatty acids, fat soluble vitamins).

Milk Proteins

Introduction and Review

The primary structure of proteins consists of a polypeptide chain of amino acids residues joined together by peptide linkages, which may also be cross-linked by disulphide bridges. Amino acids contain both a weakly basic amino group, and a weakly acid carboxyl group both connected to a hydrocarbon chain, which is unique to different amino acids.

The three-dimensional organization of proteins, or conformation, also involves secondary, tertiary, and quaternary structures. The secondary structure refers to the spatial arrangement of amino acid residues that are near one another in the linear sequence. The alpha-helix and ß-pleated sheat are examples of secondary structures arising from regular and periodic steric relationships.

The tertiary structure refers to the spatial arrangement of amino acid residues that are far apart in the linear sequence, giving rise to further coiling and folding. If the protein is tightly coiled and folded into a somewhat spherical shape, it is called a globular protein. If the protein consists of long polypeptide chains which are intermolecularly linked, they are called fibrous proteins. Quaternary structure occurs when proteins with two or more polypeptide chain subunits are associated.

Caseins

The casein content of milk represents about 80% of milk proteins. The principal casein fractions are alpha(s1) and alpha(s2)-caseins, ß-casein, and kappa-casein. The distinguishing property of all caseins is their low solubility at pH 4.6. The common compositional factor is that caseins are conjugated proteins, most with phosphate group(s) esterified to serine residues. These phosphate groups are important to the structure of the casein micelle. Calcium binding by the individual caseins is proportional to the phosphate content.

The conformation of caseins is much like that of denatured globular proteins. The high number of proline residues in caseins causes particular bending of the protein chain and inhibits the formation of close-packed, ordered secondary structures. Caseins contain no disulfide bonds. As well, the lack of tertiary structure accounts for the stability of caseins against heat denaturation because there is very little structure to unfold. Without a tertiary structure there is considerable exposure of hydrophobic residues. This results in strong association reactions of the caseins and renders them insoluble in water.

Within the group of caseins, there are several distinguishing features based on their charge distribution and sensitivity to calcium precipitation:

Alpha(s1)-casein: (molecular weight 23,000; 199 residues, 17 proline residues) Two hydrophobic regions, containing all the proline residues, separated by a polar region, which contains all but one of eight phosphate groups. It can be precipitated at very low levels of calcium.

Alpha(s2)-casein: (molecular weight 25,000; 207 residues, 10 prolines) Concentrated negative charges near N-terminus and positive charges near C-terminus. It can also be precipitated at very low levels of calcium.

B-casein: (molecular weight 24,000; 209 residues, 35 prolines) Highly charged N-terminal region and a hydrophobic C-terminal region. Very amphiphilic protein acts like a detergent molecule. Self association is temperature dependant; will form a large polymer at 20° C but not at 4° C. Less sensitive to calcium precipitation.

Kappa-casein: (molecular weight 19,000; 169 residues, 20 prolines) Very resistant to calcium precipitation, stabilizing other caseins. Rennet cleavage at the Phe105-Met106 bond eliminates the stabilizing ability, leaving a hydrophobic portion, para-kappa-casein, and a hydrophilic portion called kappa-casein glycomacropeptide (GMP), or more accurately, caseinomacropeptide (CMP).

Structure: The Casein Micelle

Most, but not all, of the casein proteins exist in a colloidal particle known as the casein micelle. Its biological function is to carry large amounts of highly insoluble CaP to mammalian young in liquid form and to form a clot in the stomach for more efficient nutrition. Besides casein protein, calcium and phosphate, the micelle also contains citrate, minor ions, lipase and plasmin enzymes, and entrapped milk serum. These micelles are rather porous structures, occupying about 4 ml/ g and 6-12% of the total volume fraction of milk. The "casein sub-micelle" model has been prominent for the last several years, and is illustrated and described with the following link, but there is not universal acceptance of this model, and mounting research evidence to suggest that there is not a defined sub-micellar structure to the micelle at all. Another model of a more open structure is also defined with the following link.

In the submicelle model, it is thought that there are small aggregates of whole casein, containing 10 to 100 casein molecules, called submicelles. It is thought that there are two different kinds of submicelle; with and without kappa-casein. These submicelles contain a hydrophobic core and are covered by a hydrophilic coat which is at least partly comprised of the polar moieties of kappa-casein. The hydrophilic CMP of the kappa-casein exists as a flexible hair.

The open model also suggests there are more dense and less dense regions within the midelle, but there is less of a well-defined structure. In this model, calcium phosphate nanoclusters bind caseins and provide for the differences in density within the casein micelle.

Colloidal calcium phosphate (CCP) acts as a cement between the hundreds or even thousands of submicelles that form the casein micelle. Binding may be covalent or electrostatic. Submicelles rich in kappa-casein occupy a surface position, whereas those with less are buried in the interior. The resulting hairy layer, at least 7 nm thick, acts to prohibit further aggregation of submicelles by steric repulsion. The casein micelles are not static; there are three dynamic equilibria between the micelle and its surroundings:

- the free casein molecules and submicelles
- the free submicelles and micelles
- the dissoved colloidal calcium and phosphate.

The following factors must be considered when assessing the stability of the casein micelle:

Role of Ca++: More than 90% of the calcium content of skim milk is associated in some way or another with the casein micelle. The removal of Ca++ leads to reversible dissociation of ß-casein without micellular disintegration. The addition of Ca++ leads to aggregation.

H Bonding: Some occurs between the individual caseins in the micelle but not much because there is no secondary structure in casein proteins.

Disulphide Bonds: alpha(s1) and ß-caseins do not have any cysteine residues. If any S-S bonds occur within the micelle, they are not the driving force for stabilization.

Hydrophobic Interactions: Caseins are among the most hydrophobic proteins and there is some evidence to suggest they play a role in the stability of the micelle. It must be remembered that hydrophobic interactions are very temperature sensitive.

Electrostatic Interactions: Some of the subunit interactions may be the result of ionic bonding, but the overall micellar structure is very loose and open.

Van der Waals Forces: No sucess in relating these forces to micellular stability.

Steric stabilization: As already noted, the hairy layer interferes with interparticle approach.

There are several factors that will affect the stability of the casein micelle system:

Salt content: affects the calcium activity in the serum and calcium phosphate content of the micelles.

pH: lowering the pH leads to dissolution of calcium phosphate until, at the isoelectric point (pH 4.6), all phosphate is dissolved and the caseins precipitate.

Temperature: at 4° C, beta-casein begins to dissociate from the micelle, at 0° C, there is no micellar aggregation; freezing produces a precipitate called cryo-casein.

Heat Treatment: whey proteins become adsorbed, altering the behaviour of the micelle.

Dehydration: by ethanol, for example, leads to aggregation of the micelles.

When two or more of these factors are applied together, the effect can also be additive.

Casein Micelle Aggregation

Caseins are able to aggregate if the surface of the micelle is reactive. The Schmidt model further illustrates this.

Although the casein micelle is fairly stable, there are four major ways in which aggregation can be induced:

1. chymosin-rennet or other proteolytic enzymes as in Cheese manufacturing
2. acid
3. heat
4. age gelation.

Enzyme Coagulation

Chymosin, or rennet, is most often used for enzyme coagulation. During the primary stage, rennet cleaves the Phe(105)-Met(106) linkage of kappa-casein resulting in the formation of the soluble CMP which diffuses away from the micelle and para-kappa-casein, a distinctly hydrophobic peptide that remains on the micelle. The patch or reactive site, as illustrated in the above image, that is left on the micelles after enzymatic cleavage is necessary before aggregation of the paracasein micelles can begin.

During the secondary stage, the micelles aggregate. This is due to the loss of steric repulsion of the kappa-casein as well as the loss of electrostatic repulsion due to the decrease in pH. As the pH approaches its isoelectric point (pH 4.6), the caseins aggregate. The casein micelles also have a strong tendency to aggregate because of hydrophobic interactions. Calcium assists coagulation by creating isoelctric conditions and by acting as a bridge between micelles. The temperature at the time of coagulation is very important to both the primary and secondary stages. With an increase in temperature up to 40° C, the rate of the rennet reaction increases. During the secondary stage, increased temperatures increase the hydrophobic reaction. The tertiary stage of coagulation involves the rearrangement of micelles after a gel has formed. There is a loss of paracasein identity as the milk curd firms and syneresis begins.

Acid Coagulation

Acidification causes the casein micelles to destabilize or aggregate by decreasing their electric charge to that of the isoelectric point. At the same time, the acidity of the medium increases the solubility of minerals so that organic calcium and phosphorus contained in the micelle gradually become soluble in the aqueous phase. Casein micelles disintegrate and casein precipitates. Aggregation occurs as a result of entropically driven hydrophobic interactions.

Heat

At temperatures above the boiling point casein micelles will irreversibly aggregate. On heating, the buffer capacity of milk salts change, carbon dioxide is released, organic acids are produced, and tricalcium phophate and casein phosphate may be precipitated with the release of hydrogen ions.

Age Gelation

Age gelation is an aggregation phenomenon that affects shelf-stable, sterilized dairy products, such as concentrated milk and UHT milk products. After weeks to months storage of these products, there is a sudden sharp increase in viscosity accompanied by visible gelation and irreversible aggregation of the micelles into long chains forming a three-dimensional network. The actual cause and mechanism is not yet clear, however, some theories exist:

1. Proteolytic breakdown of the casein: bacterial or native plasmin enzymes that are resistant to heat treatment may lead to the formation of a gel
2. Chemical reactions: polymerization of casein and whey proteins due to Maillard type or other chemical reactions
3. Formation of kappa-casein-ß-lactoglobulin complexes

An excellent source of information on casein micelle stability can be found in Walstra.

Whey Proteins

The proteins appearing in the supernatant of milk after precipitation at pH 4.6 are collectively called whey proteins. These globular proteins are more water soluble than caseins and are subject to heat denaturation. Native whey proteins have good gelling and whipping properties. Denaturation increases their water holding capacity. The principle fractions are ß-lactoglobulin, alpha-lactalbumin, bovine serum albumin (BSA), and immunoglobulins (Ig).

ß-Lactoglobulins: (MW-18,000; 162 residues) This group, including eight genetic variants, comprises approximately half the total whey proteins. ß-Lactoglobulin has two internal disulfide bonds and one free thiol group. The conformation includes considerable secondary structure and exists naturally as a noncovalent linked dimer. At the isoelectric point (pH 3.5 to 5.2), the dimers are further associated to octamers but at pH below 3.4, they are dissociated to monomers.

Alpha-Lactalbumins: (MW-14,000; 123 residues) These proteins contain eight cysteine groups, all involved in internal disulfide bonds, and four tryptophan residues. alpha-Lactalbumin has a highly ordered secondary structure, and a compact, spherical tertiary structure. Thermal denaturation and pH <4.0 results in the release of bound calcium.

Enzymes

Enzymes are a group of proteins that have the ability to catalyze chemical reactions and the speed of such reactions. The action of enzymes is very specific. Milk contains both indigenous and exogenous enzymes. Exogenous enzymes mainly consist of heat-stable enzymes produced by psychrotrophic bacteria: lipases, and proteinases. There are many indigenous enzymes that have been isolated from milk. The most significant group are the hydrolases:

- lipoprotein lipase
- plasmin
- alkaline phosphatase.

Lipoprotein lipase (LPL): A lipase enzyme splits fats into glycerol and free fatty acids. This enzyme is found mainly in the plasma in association with casein micelles. The milkfat is protected from its action by the FGM. If the FGM has been damaged, or if certain cofactors (blood serum lipoproteins) are present, the LPL is able to attack the lipoproteins of the FGM. Lipolysis may be caused in this way.

Plasmin: Plasmin is a proteolytic enzyme; it splits proteins. Plasmin attacks both ß-casein and alpha(s2)-casein. It is very heat stable and responsible for the development of bitterness in pasteurized milk and UHT processed milk. It may also play a role in the ripening and flavour development of certain cheeses, such as Swiss cheese.

Alkaline phosphatase: Phosphatase enzymes are able to split specific phosporic acid esters into phosphoric acid and the related alcohols. Unlike most milk enzymes, it has a pH and temperature optima differing from physiological values; pH of 9.8. The enzyme is destroyed by minimum pasteurization temperatures, therefore, a phosphatase test can be done to ensure proper pasteurization.

Lactose

Lactose is a disaccharide (2 sugars) made up of glucose and galactose (which are both monosaccharides). Because of the anomeric carbon on the right side of the structure below, lactose can exist as two isomers, alpha, as shown, or beta, in which the hydroxyl on the anomeric carbon would point up on the ring structure shown below.

It comprises 4.8 to 5.2% of milk, 52% of milk SNF, and 70% of whey solids. It is not as sweet as sucrose. When lactose is hydrolyzed by ß-D-galactosidase (lactase), an enzyme that splits these monosaccharides, the result is increased sweetness, and depressed freezing point.

One of its most important functions is its utilization as a fermentation substrate. Lactic acid bacteria produce lactic acid from lactose, which is the beginning of many fermented dairy products. Because of their ability to metabolize lactose, they have a competitive advantage over many pathogenic and spoilage organisms.

Some people suffer from lactose intolerance; they lack the lactase enzyme, hence they cannot digest lactose, or dairy products containing lactose. Crystallization of lactose occurs in an alpha form which commonly takes a tomahawk shape. This results in the defect called sandiness. Lactose is relatively insoluble which is a problem in many dairy products, ice cream, sweetened condensed milk. In addition to lactose, fresh milk contains other carbohydrates in small amounts, including glucose, galactose, and oligosaccharides.

Vitamins

Vitamins are organic substances essential for many life processes. Milk includes fat soluble vitamins A, D, E, and K. Vitamin A is derived from retinol and ß-carotene. Because milk is an important source of dietary vitamin A, fat reduced products which have lost vitamin A with the fat, are required to supplement the product with vitamin A.

Milk is also an important source of dietary water soluble vitamins:

- B1-thiamine
- B2-riboflavin
- B6-pyridoxine
- B12-cyanocobalamin
- niacin
- pantothenic acid.

There is also a small amount of vitamin C (ascorbic acid) present in raw milk but it is an insignificant amount relative to human needs and is quite heat-labile: about 20% is destroyed by pasteurization.

The vitamin content of fresh milk is given below:

Vitamin	*Contents per litre*
A (ug RE)	400
D (IU)	40
E (ug)	1000
K (ug)	50
B1 (ug)	450
B2 (ug)	1750
Niacin (ug)	900
B6 (ug)	500
Pantothenic acid (ug)	3500
Biotin (ug)	35
Folic acid (ug)	55
B12 (ug)	4.5
C (mg)	20

Minerals

All 22 minerals considered to be essential to the human diet are present in milk. These include three families of salts:

1. *Sodium (Na), Potassium (K) and Chloride (Cl):These* free *ions are negatively correlated to lactose to maintain osmotic equilibrium of milk with blood.*
2. *Calcium (Ca), Magnesium (Mg), Inorganic Phosphorous (P(i)), and Citrate: This group consists of 2/3 of the Ca, 1/3 of the Mg, 1/2 of the P(i), and less than 1/10 of the citrate in* colloidal *(nondiffusible) form and present in the casein micelle.*
3. Diffusible *salts of Ca, Mg, citrate, and phosphate: These salts are very pH dependent and contribute to the overall acid-base equilibrium of milk.*

The mineral content of fresh milk is given below. A wide range is shown, based on variability due to intrinsic (lactogenesis in the secretory cell of individual animals) and extrinsic factors (mainly feed and season/climate adaptation, but also contaminants).

This information is consolidated from C. D. Hunt and F. H. Nielsen, P.L.H. and P.F. Fox, 2009. Advanced Dairy Chemistry, Vol. 3, Lactose, water, salts and minor constituents, 3rd edn.,

Mineral	*Content per litre*
Sodium (mg)	250-640
Potassium (mg)	1100-1500
Chloride (mg)	800-1200
Calcium (mg)	1100-1300
Magnesium (mg)	70-140
Phosphorus (mg)	800-1000
Iron (ug)	100-700
Zinc (ug)	2500-7000
Copper (ug)	100-350
Manganese (ug)	10-50
Iodine (ug)	50-600
Fluoride (ug)	20-80
Selenium (ug)	20-40
Cobalt (ug)	0.5-1.3
Chromium (ug)	0.5-20
Molybdenum (ug)	20-100
Nickel (ug)	0-50
Arsenic (ug)	20-60
Aluminum (ug)	50-1600
Lead (ug)	tr-20

Springer, pp. 392-8; and from, Cristina Sola-Larranaga; Inigo Navarro-Blasco. 2009. Chemometric analysis of minerals and trace elements in raw cow milk from the community of Navarra, Spain. Food Chemistry, 112 (1), 189-196.

Physical Properties

Density

The density of milk and milk products is used for the following;

- to convert volume into mass and vice versa
- to estimate the solids content
- to calculate other physical properties (e.g. kinematic viscosity).

Density, the mass of a certain quantity of material divided by its volume, is dependant on the following:

- temperature at the time of measurement
- temperature history of the material

- composition of the material (especially the fat content)
- inclusion of air (a complication with more viscous products).

With all of this in mind, the density of milk varies within the range of 1027 to 1033 kg m(-3) at 20° C.

Viscosity

Viscosity of milk and milk products is important in determining the following:

- the rate of creaming
- rates of mass and heat transfer
- the flow conditions in dairy processes.

Milk and skim milk, excepting cooled raw milk, exhibit Newtonian behaviour, in which the viscosity is independent of the rate of shear. The viscosity of these products depends on the following:

- Temperature:
  - — cooler temperatures increase viscosity due to the increased voluminosity of casein micelles
  - — temperatures above 65° C increase viscosity due to the denaturation of whey proteins
- pH: an increase or decrease in pH of milk also causes an increase in casein micelle voluminosity

Cooled raw milk and cream exhibit non-Newtonian behaviour in which the viscosity is dependant on the shear rate. Agitation may cause partial coalescence of the fat globules (partial churning) which increases viscocity. Fat globules that have under gone cold agglutination, may be dispersed due to agitation, causing a decrease in viscosity.

Freezing Point

Freezing point depression is a colligative property which is determined by the molarity of solutes rather than by the percentage by weight or volume. In the dairy industry, freezing point of milk is mainly used to determine added water but it can also been used to determine lactose content in milk, estimate whey powder contents in skim milk powder, and to determine water activity of cheese. The freezing point of milk is usually in the range of-0.512 to-0.550° C with an average of about-0.522° C. Correct interpretation of freezing point data with respect to added water depends on a good understanding of the factors affecting freezing point depression. With respect to interpretation of freezing points for added water determination, the most significant variables are the nutritional status of the herd and the access to water. Under feeding causes increased freezing points. Large temporary increases in freezing point occur after consumption of large amounts of water because milk is iso-osmotic with blood. The primary sources of non-intentional added water in milk are residual rinse water and condensation in the milking system.

Acid-Base Equilibria

Both titratable acidity and pH are used to measure milk acidity. The pH of milk at 25° C normally varies within a relatively narrow range of 6.5 to 6.7. The normal range for titratable acidity of herd milks is 13 to 20 mmol/L. Because of the large inherent variation, the measure of titratable acidity has little practical value except to measure changes in acidity (eg., during lactic fermentation) and even for this purpose, pH is a better measurement.

There are many components in milk which provide a buffering action. The major buffering groups of milk are caseins and phosphate.

Optical Properties

Optical properties provide the basis for many rapid, indirect methods of analysis such as proximate analysis by infrared absorbency or light scattering. Optical properties also determine the appearance of milk and milk products. Light scattering by fat globules and casein micelles causes milk to appear turbid and opaque. Light scattering occurs when the wave length of light is near the same magnitude as the particle. Thus, smaller particles scatter light of shorter wavelengths. Skim milk appears slightly blue because casein micelles scatter the shorter wavelengths of visible light (blue) more than the red. The carotenoid precursor of vitamin A, ß-carotene, contained in milk fat, is responsible for the 'creamy' colour of milk. Riboflavin imparts a greenish colour to whey.

Refractive index (RI) is normally determined at 20° C with the D line of the sodium spectrum. The refractive index of milk is 1.3440 to 1.3485 and can be used to estimate total solids.

4

Efficiency and Cost of Milk Production

Milk and Dairy Products: Production and Processing Costs

In "liquid milk equivalent" international milk prices are less than half the price paid for domestic liquid milk in many Third World cities. Though the international milk market may fluctuate in the short term, it is likely to remain low in the years to come. From a strictly economic standpoint, therefore, it might not seem reasonable to encourage Third World countries to develop dairy programmes. However, development potential and specifics do vary considerably from country to country and from one region to the next. An individual government may have its own social or political reasons for wishing to develop milk production.

In point of fact:

- The sale of milk provides extra income on a regular basis to rural people who often have just enough to get by. Milk sales also improve the nutritional status of rural populations, though not necessarily the urban needy. Milk is also one of the few agricultural products which can be supplied and marketed regularly by non-landowners. A dairy plant creates jobs in rural areas, and thus helps check urban migration.
- Milk and milk products are thought of as staple foods in Europe. Favourable production conditions mean that every one can consume large amounts of milk and dairy products. For many of the world's peoples, however, even though milk plays no role in the diet, if balanced food is available in sufficient quantities, neither health nor welfare should be affected by the

lack of milk. Recommended Third World dairy policies fall somewhere between these two extremes. As FAO Agricultural Paper No. 89 (FAO 1973) shows, milk protein can enhance plant protein. In a diet combining several foods, a compensatory balance is established among the various sources of proteins. Animal proteins supply essential amino acids, plant proteins can then economically top up the total supply.

- Many Third World countries find it hard to right the balance of payments: imported dairy products are one foreign exchange savings which can be effected by a dairy development policy. Moreover, international prices may go up in the medium or long run, placing a country which is not in a position to partially substitute imports by local production at a disadvantage.

Many Third World countries have decided to develop national dairy production for the above reasons. A certain amount of protectionism is necessary to ensure success and avert unfair competition between producer country-subsidized dairy imports and domestic production.

The government needs to tax dairy imports, particularly powdered, concentrated or evaporated milk products. The revenues are used to finance dairy development projects or to boost producer prices (subsidizing some major components of production costs). For the consumer, the outcome of these measures is high-priced milk and dairy products. In many developing countries, the cost of one litre of milk is equivalent to more than 5 percent of the median daily income-in some cases as much as 60 percent.

Consumer milk prices have a great psychological impact. Many governments are tempted to try to regulate milk prices to keep the urban population happy, but milk producers' associations often have great economical and political strength. This leaves the government very little room for manoeuvre, caught as it is between producers and consumers. The dangers inherent in a failure to take appropriate measures are:

- Excessively low processing industry profits, which bar re-investment and, in a few years, leave the plant with obsolete equipment thus driving up its processing costs.
- Rock-bottom producer prices, which discourage production development, encourage the rise of a parallel market and the sale of adulterated, unpasteurized milk.

To act with discernment, a government therefore needs to know milk production costs in its major milk-producing regions, as well as the costs of processing and marketing milk and of the major dairy products. Getting such data is, in practice, a highly complex matter, and the figures tend to be approximative.

Production costs do in fact fluctuate greatly in accordance with:

- the region under consideration (depending on soils, climate, etc.);
- herd size;
- level of technical skills;
- whether the enterprise is multi-purpose and how much area is available.

Likewise, processing costs are frequently not known as most dairies simply list all manufacturers' gross profits without itemizing the real processing cost of each of the individual products manufactured. In addition, these costs are often too high because dairy plants have been overequipped and depreciation is excessive for the volumes actually processed. Frequently, the processing equipment does not produce the expected yields due to shortages of spare parts and inadequate maintenance. National dairy policy-makers will therefore require the most detailed information possible on the various cost prices. They will also have to consider such concepts as added value, which is connected to the social function of domestic dairy production.

Added value represents the difference between the value of output over a specific period of time and the value of the goods and services purchased abroad and used to produce this output during this same period. Added value during a given period can be measured by adding up the following items: personnel costs, duties and taxes, financial costs, depreciation, profit.

In a way, added value is a measure of an enterprise's contribution to the national wealth. It is a much better reflection of actual business activity than turnover, and is the figure which expresses the purpose of the business for time and place, e.g. comparisons between different years and different enterprises.

Ratios such as, for example:

$$\frac{\text{Personnel costs}}{\text{Added value}} \qquad \frac{\text{Depreciation}}{\text{Added value}}$$

It possible to measure the respective share of staff and equipment in the wealth generated by the enterprise.

An increase in added value has the following effects:

- An increase in the number of jobs generated by the enterprise. This is of particular interest to the nation as these jobs are created in rural areas where unemployment is frequently chronic.
- An enhancement of technical skills in milk production and/or processing.

Greater added value for a specific dairy firm usually means greater productivity and hence greater competitivity. If national productivity goes up, the country becomes less vulnerable to depressed international prices for milk and dairy products. Within the industry, greater added value means more sophisticated products-or enhanced values for plant by-products.

- Reduced milk or dairy product imports or, conversely, increased exports.

Estimation and Control of Cost Prices

A knowledge of cost prices is indispensable in running a firm. They are at the root not only of sale prices but also of the entire commercial and industrial policy of the business.

A business' prime objective is to ensure maximum profitability, bearing in mind production capacity or the potential to supply raw materials.

There are "choices" to be made-short-term and long-term choices and decisions to be taken-and all are based on a prior analysis of costs and cost prices.

Once the decisions have been made, it is essential to see that they are carried out in accordance with objectives and projections. This is the pivotal problem of cost control.

The concept of cost price is always somewhat arbitrary and conventional no matter how much care is taken to work out allocations and invoicing.

These calculations are of course based on certain assumptions:

- assumed distribution
- assumed activity.

The Point of Cost Prices

Any concern needs to know the final costs of a given type of product before deciding to develop and distribute one product rather than another.

It has a general accounting section which reports on overall plant performance. There are, however, two major drawbacks:

- returns (loss or profit) become apparent at the end of the fiscal year (too late);
- the presentation is not conducive to a cost breakdown by product type-nor does it show eventual losses of raw materials, or of steam, electricity, chemical products, packing materials, etc., or under-use of material and labour.

The data from the general accounting section are indispensable for the tax auditor and for people external to the firm such as shareholders and bankers but they are insufficiently detailed for engineers and administrators. General accounting must therefore be paralleled by cost control.

The establishment of a system for calculating manufacturing costs should enable a business to:

Determine Sale Prices

Before launching a new manufacturing line, perhaps making fresh investment, the head of the concern needs to know whether the market situation and the competition are going to allow him to sell at a price which fully covers his expenditure.

The determination of a sale price or the decision to launch a manufacturing line is therefore based on the cost price forecast.

It is not enough to know that the sale price is higher than the cost price forecast. Since the goal is also maximum profitability, it is preferable to produce and sell articles which bring the highest return. This all requires a thorough, systematic analysis of manufacturing costs and cost prices to determine just how firm are the assumed returns-particularly those from assumed activities.

Cost Control

The decision to manufacture or to sell is preceded by a cost breakdown of the manufacturing cost forecast. It is therefore sensible, indeed probably essential to smooth operations, to ensure that forecasts will be realistic and to note eventual gaps between true costs and pre-established costs-as these normally have an impact on returns.

Basically, the audit ascertains such gaps so as to:

- detect anomalies
- identify the cause of such anomalies and eventual responsibilities
- take action to remedy losses and improve performance
- ascertain production gains due to new techniques or procedures, to improve future forecasts.

An audit of the fixed costs of periodic (usually annual) budgets does not require the same expenditure of effort as an audit of such variable costs as raw materials, labour, and activity-specific manufacturing costs.

Investment Policy

The decision to purchase or replace equipment is usually preceded by a pre-investment profitability study. Such a study investigates the impact of the new equipment on costs and cost prices.

Evaluating Assets

A stock inventory-the stock is usually assessed at cost price-is a prerequisite to totting up fiscal year returns. Cost price calculations are accordingly very important in working out returns.

The articles in stock (cheeses in the process of being cured) have not all been purchased or manufactured at the same time under the same cost price conditions. Calculations are based on assumptions derived from prior events, thus leaving a wide margin of uncertainty.

Evaluations of assets can vary considerably in accordance with prior assumptions and so an accurate assessment demands the application of consistent rules.

A Greater Interest in Work

One important advantage of identifying cost prices and cost controls is that this kind of calculation is worked out at the section chief level. In other words, the head of the cheese section is responsible for the variable charges attributable to the cheese section; the head of the laboratory is responsible for laboratory analyses and quality control costs.

The system is meant to make each individual feel responsible for a share of the overall cost and make the work more interesting and stimulating. A clearly-spelled-out policy on identifying costs and on management audits is a business imperative.

Classification of Charges

Direct and Indirect charges

Charges may be:

- direct (the term allocation is reserved for direct charges) or
- indirect. Here we use the term charging; charging is preceded by distribution.
  a. Direct charges: Direct charges are those which can be unambiguously identified and attributed to a cost or a cost price. They may be fixed or variable.
  b. Indirect charges: There are also other charges which cannot easily be attributed to some activity of the enterprise, and which are very difficult or impossible to identify at this level. Administrative (management) costs, for example, fall into this category.

Charges which have to be distributed somewhat arbitrarily among the various cost prices and manufacturing costs are categorized as indirect charges. Such charges most often correspond to fixed or structural charges and are sometimes termed "general expenses".

Fixed and Variable Charges

If we look at costs in connection with production we see that some costs vary with the volume of production. These are called variable costs (e.g.: quantity of milk, production workforce). Others are constant. These are called fixed costs, e.g.: rent, executives' salaries, etc.

A closer look at the question prompts the following remarks:

1. Virtually all of the so-called fixed charges are actually variable, but the coefficient of variation is not the same as for the so-called variable charges.
2. The so-called variable charges do not all vary in accordance with the same factor of production; some vary in accordance with machine running time.
3. Variable charges differ from fixed charges in that they only arise as and when the undertaking is actually producing. They represent activity costs or perhaps operational expenses whereas fixed charges are structural costs.

As a base minimum, sale prices must cover variable costs and leave some margin. The contribution of each product to the coverage of fixed charges differs in accordance with sales potential but structural costs must in any case be absorbed by overall production for a given period.

Note 1

Write-off or depreciation is the loss in value of goods over time or with use. These two essential factors of depreciation should include a third: technological ageing (obsolescence) in which depreciation is part of the manufacturing cost. Neither land nor animals are written off in a dairy production unit as these goods do not drop in value over time. But depreciation also plays a role in the financial management of the enterprise. It is the main source of self-financing to the extent that write-off is tax-deductible. As a result, tax-deductible write-off is virtually never linked to depreciation in the strict sense of the word.

The write-offs to keep in mind in calculating manufacturing costs, frequently termed "technical" or "industrial" depreciations, are calculated in terms of the probable and real use to which investments will be put, and often differ greatly from accountable, fiscal depreciations.

Depreciation varies in length from one plant to the next and even more from one country to the next. In civil engineering terms, the usual length of depreciation is from 15 to 20 years. Twenty years can be considered a top figure for industrial buildings which depreciate faster than, say, office buildings. Building maintenance costs are included under variable costs when the amounts of money involved are small. Major expenses are earmarked and amortized.

As for equipment, many firms make a distinction between static machinery and machines which have moving, electrical, electro-mechanical or electronic parts. In the first category, which includes fermenting rooms and heat-exchangers, the period of depreciation usually varies from 10 to 15 years. For the second category, which includes pumps, and automization, the period of depreciation is shorter, usually from 3 to 8 years.

There are three main ways of calculating depreciation:

1. Straight-line depreciation. The purchase cost is distributed equally over a given number of years.
2. Sliding-scale fiscal depreciation. This is a depreciation method of fiscal origin whereby investment is favoured by applying a high rate of depreciation for the first years. The first period

runs from the day of purchase to when the rate dips below the straight-line rate. During the second period, when the annuity drops below the annuity calculated by dividing the residual value by the number of years remaining, the latter figure is used.

3. An updated value of the good can form the basis of the depreciation rate, which has the effect of not substantially modifying cost prices when a fully amortized good is replaced, the cost of repurchase value generally being much higher than the original cost.

Note 2

"Auxiliary charges", though ignored by general accounting, may be taken into account in calculating cost price.

By "auxiliary charges" is meant the battery of charges for factors of production wholly owned by the head of the enterprise or exploitation, and made available to it by him.

This involves, in particular:

- Remuneration for the labour of the owner and his family (manual and perhaps managerial work).

Such remuneration must be calculated as a lump sum on the basis of the remuneration which would have been granted to equally qualified employees.

- Conventional remuneration of the "owner capital" of the enterprise.

In the case of milk production costs, rather than remunerating all of the owner capital at one time in one lump sum, it is preferable to pay off the two components of this item singly, i.e.:

- Net farm assets, i.e. the working capital less any debts which may have been contracted to finance it.

The sum is paid off by applying a lump sum rate of interest corresponding to the average rate on eventual loans to finance working capital.

- Net land assets, i.e. the total land capital, less eventual debts to finance it.

The conventional remuneration of land assets, more commonly called the rental value, is calculated by deducting the land charges already accounted for in the books from the amount of rent which could normally be anticipated from the land. This would include:

- Land taxes and duties
- Such land maintenance expenses as drainage
- The write-off for land improvements
- The financial charges on eventual loans to finance the purchase of land.

These charges are incorporated for maximum comparability of manufacturing costs and cost prices, particularly to compare two enterprises with different legal structures. In any case these charges can only be estimated by inference and are inevitably partly subjective.

Cost Centres

Consider that in the average enterprise:

- several types of products are produced;
- several different functions are performed e.g. processing, administration, purchases, etc., each of which in turn can be divided into sub-functions wherein a specific job is performed. For instance, milk processing may be subdivided into reception, pasteurization, standardization, and packaging.

The different functions of an enterprise from which its global costs can be figured are called "cost centres", or sections. Conventionally, these cost centres correspond to real, natural divisions of the enterprise and are headed by a physical person. Cost forecasts and a knowledge of the operating costs of each cost centre are basic to management audits. Dividing an enterprise into "cost centres" has two advantages:

a. It facilitates the distribution of indirect costs and the determination of the overall cost, as each cost centre behaves like a small enterprise.

b. It helps pinpoint responsibilities and differences between projected and actual figures.

Cost centres are generally divided into three broad categories:

a. Main centres-activities directly involved with the production and distribution phases, such as milk reception or pasteurization.

b. Auxiliary centres-which unlike the preceding, work for the benefit of other sections. An auxiliary section may well work for other auxiliary sections-producing steam or electricity, for example. Such costs need to be broken down among the various main sections.

c. Structural costs centres. These costs correspond to functions whose costs are independent of production (management, the accounts department, etc.).

Ascertaining and Analysing Differentials

There are three major classes of differentials or deviations:

- on goods
- on direct labour
- cost differentials.

On Goods

These are due to consumption (the audit of goods) and prices (value differentials).

Quantitative Differentials

Quantitative differentials are checked daily by tallying the quantities of milk received and the plant output. Milk reception and manufacturing slips provide the basis for the daily audit. Quantitative differentials should also be checked regularly for all goods in store and used at some stage of processing.

Quantitative differentials can usually be attributed to some "lapse" of one or several services;

- processing (carelessness, misuse of equipment);
- maintenance (equipment in poor running order);
- management audit (forecasting error).

Exceptionally, there may be differentials of external origin which cannot be laid at the feet of some subsection of the enterprise. Examples are qualitative variations in the milk received, substantial variations in atmospheric conditions, and so forth.

Value Differentials

Value differentials will be noted by the milk supply service or by the purchasing service for all articles in store.

Value differentials may be either internal or external.

- internal
- Purchasing Service-unfavourable sales conditions or excessive maintenance charges.

- Processing Section-abnormal consumption forced the service to resupply at higher cost.
- Commercial Section-the volume of sales did not match forecasts, and the purchasing programme had to be modified.
- external.

Price fluctuations due to monetary depreciation, changing international rates, and the like.

Manpower Differentials

Quantitative differentials: These are due to downtime from stoppages (mechanical breakdowns, shortage of materials) or poorly-organized work. They are basically internal and virtually any section or service may be involved.

Value differentials: These can be picked up from the wage rolls, and may be of internal or external origin.

Cost Differentials

A distinction may be made between:

a. variable cost differentials, usually perceived in quantity and in value.

Actual consumption may be picked up from:

- meter readings of consumption (e.g. water, electricity);
- issue vouchers from stores or current purchases, taking into account inventory differences such as consumables;
- charges accounted for during the current period, e.g. maintenance expenses.

b. Fixed cost differentials.

This is the difference, which usually shows up in the books, between budgeted costs and actual ascertained costs.

Cost differentials are broken down into:

- budget differential
- operation or activity differential
- yield differential.

The differential analysis is made on the basis of the "adapted" budget in order to eliminate the activity differential at an early stage.

The budgetary differential can be analysed in variable component differentials by distinguishing quantitative and value differentials and fixed charge component differentials for each one.

Generally speaking, the budget differential expresses a variation in expenses due to a weakness in implementation or an error in forecasting.

Activity Differential : This is a way of measuring the impact of the activity on the coverage of structural charges.

If there is some sub-activity, the fixed costs are never completely covered. The difference between budgeted fixed costs and absorbed costs represents the cost of the sub-activity.

On the other hand, if real activity exceeds planned activity, there is excess coverage which is expressed as an over-activity bonus.

The activity index is the ratio of real to planned activity;

$$a_I = \frac{\text{Real activity}}{\text{Planned activity}}$$

The responsibility for the activity differential lies at the very top, with management, who uses the basic data to set business policies, chooses among the various possible options, and sets activity targets.

Yield differential : The yield differential is the difference between the production expected from real activity and what is actually produced. The productivity index is the ratio between real production and "normal" or expected production

$$p_I = \frac{\text{Real production}}{\text{Normal production}}$$

Comments on the above Diagram

The forecasts established a given production level P_1. A regular budget was therefore determined for this estimated activity, including fixed and variable costs. For this level of activity the enterprise will cover all costs and make a profit EF. The forecasts also show that if the only differentials are activity differentials, the break-even point (no profits, no losses) of the enterprise will be reached at production P_4. True production was only P_2 and total costs OA. In this case the budgetary differential is a malus-AB.

The quantity of milk received led to production forecast P_3. True production, however, was P_2. This made possible a bonus on the yield differential-CD.

The milk processing service chief will be responsible only for malus AB and bonus CD. On the other hand, malus BC will be attributed to management who made the wrong assumptions concerning activity.

Estimated profits EF thus ended up as estimated loss AG. Actual losses will depend on actual sales.

If production exceeds P_1 (and assuming the plant turns out only one product) there are two possibilities:

a. Consider that profit is higher than estimated, as in the preceding case.

b. Consider profit and estimated profit to be equal and make it play a greater part in covering fixed costs.

Profit is shown by the ratio:

$$\text{Profit} = P\,(S\text{-}C_v)\text{-}C_f$$

in which:

P = True production

S = Unit sale price

C_v = Variable unit costs

C_f = Fixed costs

Break-even production (BE) is shown by the ratio:

$$B_E = \frac{C_F}{S - C_V} = \frac{C_F}{UMC_V}$$

with UMCv equalling unit margin (profit) on variable costs.

1. Action on fixed costs. Any substantial percentual reduction of fixed costs entails an equal reduction of break-even production.

   Irrespective of the amounts sold, profits rise to the extent that fixed costs fall. Saving on fixed costs is therefore real, immediate saving, unlike economies on variable costs which materialize only in accordance with sales.

2. Action on variable costs. A reduction in variable unit costs lowers the break-even point. The impact on break-even production increases as the initial margin on variable costs narrows.

   The impact on profit, of course, depends on the volume of sales.

3. Action on sale prices. A rise in sale prices will lower the break-even point. There again, the impact on break-even production increases as the initial margin on variable costs narrows. The impact on profit is also directly affected by volume of sales.

Setting up a budgetary control procedure (management audit)

The purpose of a management audit is to:

- prepare estimates (budgets);
- periodically compare performance to forecasts;
- analyse differentials and zoom in on anomalies.

Budgetary techniques are complex operations which the enterprise should introduce cautiously. There are certain pre-conditions:

1. Appointment of an officer in charge. Someone needs to be put in charge of budget preparation, monitoring and analysis. The problem is to find someone who has both the necessary technical background and the book-keeping skills. The management auditor is usually a staff member working directly under the top management. In small-and medium-size businesses, it is not always possible to create a post just for this purpose and so the job is done by the head accountant, the technical director or the head of the business himself.
2. Initial organization and structuration of the enterprise. A clearly spelled-out organizational chart designating responsibilities and outlining the organization of work stations is one essential pre-condition for setting up the system. In particular, the enterprise must be broken down into cost centres once the technical context has been reviewed. Each cost centre should be headed by one person who will help to prepare and subsequently modify budgets.
3. Paving the way. In order to create a favourable climate, the people in charge will need to have some prior information. The notion of auditing is not easily and readily accepted. To secure the active cooperation of all involved, the purpose of the exercise must first be explained.

Once this has been done, it would be a good idea to plan a review of how to implement the system, detailing:

- how forecasts and estimates are made;
- which items are to be audited;
- the procedure to be used;
- how the information is to be disseminated.

It is prudent to envisage a "breaking in" period as it takes some time to learn to make budgetary estimates-from a few months to a few years depending on the business enterprise. This is a precaution against the transmission of unusable or erroneous data. The system may also be introduced in successive stages.

Milk Production Costs

General Considerations

Milk production costs differ from country to country, from one producer area to the next, and among production economies. Determining production costs is therefore a delicate matter and should be done on a case-by-case basis. Extrapolating from the figures obtained can only produce approximative results.

A certain amount of data are needed on the context of the dairy farm or business to be costed:

- geographical context
- economic climate
- available farmland
- staff
- products other than milk
- herd size and composition
- fodder system
- animal housing and milking systems.

The gross margin produced by the activity "milk production" can be expressed in terms of:

- number of dairy cows
- ha of fodder area
- number of labourers
- available capital.

The choice of which ratio to use will depend on the individual farm situation, i.e. on its own limiting factors.

A livestock producer with a fixed installation which he does not wish to modify will be more apt to select "number of dairy cows". He may indeed have a fixed amount of labour to work with the herd, for which reason he wishes to maintain a constant number of animals.

A livestock producer with a limited amount of land is more apt to choose "hectares of fodder area", though he may modify the herd size if he is willing to make new investments.

If available labour is the limiting factor, the farmer will try to get the maximum gross margin per labour unit available on the farm.

Available capital is the limiting factor for many farmers. In this case gross margin will be compared to the capital invested.

If a government wishes to develop dairy production, then, it should establish a farm-gate milk price which will make dairy production more profitable in terms of the ratio chosen by the farmer than other possible farming activities in that area. The cost price of one litre of milk is therefore not the main criterion for a milk producer. He ought rather to look at the gross margin in terms of the specific limiting factor of his own farm.

What we usually see in the developing countries is that fodder cropland and dairy potential are so underexploited that any effort to intensify production and adapt resources to stock requirements initially entails reduction of costs per unit of milk produced. Logically then, by simultaneously intensifying fodder production and dairy production, the cost of producing milk begins to drop as production rises, significantly pushing up gross margin.

Specialization and intensification, however, mean boosting production at the cost of rising operating expenses (more fertilization for fodder crops; diversification of the fodder system; changes in how the livestock production system is run and in the basic livestock diet; better fodder quality; using more concentrates; adding more dairy cows and breeding better milkers, and so forth).

Cost Accounting Applied to the Agricultural Enterprise

Like any enterprise, an agricultural enterprise needs to be broken down into main headings corresponding to its specific activities.

These main headings may include:

- dairy production
- livestock production
- cereal production
- food crop production, etc.

These cost centres correspond to the different purposes the farm has set itself. It is quite rare, particularly in the developing countries, to find mono-purpose farms.

To ensure that these various sections of the farm run smoothly, the farmer also needs to develop a number of auxiliary sections:

- agricultural equipment
- repair and maintenance shop
- harnessing and treating water
- storage areas for fertilizer, veterinary stocks, plant protection products, harvests and the like.

And, like any enterprise, the farm will have structural costs which will, of course, be relatively minor for small farms. Farms, unlike most industrial or commercial enterprises, are often both a consumption unit the farmer and farm family) and a production unit. So cost accounting will need to reckon in all withdrawals in cash or in kind by the farmer to meet his/her own needs.

Agricultural economics uses the term "farm income". This is the amount remaining to the farmer and farm family after operating expenses. Such income therefore serves to remunerate the labour and capital supplied by these people.

Farm income differs from available income because part or even all of it may have been used during the year due to a variation in stocks or to investment expenditure. On the other hand, it includes costs which are not translated into actual expenditures of money such as depreciations.

In order to account for auxiliary charges, which are counted in cost accounting, we shall speak of the income of the business enterprise:

Business income= Farm income- Auxiliary charges

Auxiliary charges basically comprise:

- The rental value of land owned by the farmer. This is the potential rental value of the land which the farmer would have to pay were he forced to rent.
- The interest from capital commitments (excepting land).
- Wages not paid to the farmer and his family.

Farm income = Farm output - Operating expenses

The farm output consists of the product of the various activities on the one hand, and of such various yields as:

- subsidies
- work done for third parties
- financial proceeds
- work done by the enterprise for its own benefit.

Farm income not being the same as available income, it would be in the farmer's interest to draw up both a chart of fiscal year finances and a budgetary estimate for the following financial year. This second table will enable him to avoid a cash flow problem, indicating how much he can withdraw for his own reasons.

Unvarying rules for evaluating stocks and fixed assets must be strictly followed. Only the enterprise's own land appears on the inventory. For the first inventory, the average price of land of equal quality in the area will be taken as a base datum. Land purchased during the financial year will be inventoried at purchase value. The inventory value of buildings, equipment and installations equals the purchase value less cumulative write-off from previous financial years. Livestock will be listed under stocks, not capital.

There should be three categories of commodities in stock:

a. Supplies in stock
b. Harvest stores
c. Advances for crops. This includes all sums earmarked for future harvests such as seeds in the ground, fertilizers, and so forth.

Supplies and advances for crops are inventoried at purchase value. Harvest stores such as cereals, straw and hay are inventoried in bulk by amount at the probable market rate at the time the inventory is made. For unmarketed crops such as silage and fodder beets, the fodder unit value will be reckoned in comparison with equivalent market commodities such as hay.

The milk produced in a dairy establishment can be marketed in different ways. It may be:

- sold to a dairy
- retailed
- used for farm family consumption or part of it given to hired farm labourers
- used to feed calves
- used as the raw material for a processing unit producing pasteurized milk, cheese, yoghurt or butter.

These different marketing systems should be compared to determine the more profitable one for a given dairy farm. Cost accounting makes it possible to compare the cost price of milk with the purchase price paid by the dairy. If the dairy buys refrigerated milk, the cost of refrigeration will have to be subtracted from the price of one litre of milk to get the price of one litre from the milking room.

For retail sales, distribution costs must be added to the cost price of milk from the milking room to check whether retail sales are more profitable than selling to the dairy. Milk consumed by the farm family or farm workers will be accounted for at cost price and listed under the heading : "benefits in kind".

The cost price of milk will be compared to substitute feeds to determine which would be preferable: to save it for feeding calves or to sell the milk and buy substitute feeds.

If the calves are milk-fed, the estimated milk price can be lowered as milking costs do not have to be taken into account. Such milk will constitute an internal sale between the "milk production" section and the "livestock production" section. The price of this milk will be considered as proceeds for the "milk production" section and charged to the "livestock production" section. This internal sale system is essential for determining returns per activity or operation, but has no bearing on the overall returns of the dairy farm.

Livestock

Livestock is one of the major assets of an agricultural enterprise specializing in milk production. A set of criteria are commonly used to describe herd composition, trends, reproduction and milk production. These criteria are used for technical management of the animals and constitute useful elements for evaluating economic performance.

Herd size also needs to be compared to available fodder crop area to make sure the herd will be properly fed.

Herd Composition

The following are used to describe herd composition:

Number of Dairy Cows: This is the number of dairy cows actually present in the stable during the financial year. The presence of both milkers and dry cows, whether present all or part of the year, is counted in days and the total number of days divided by 365. The number of dairy cows present comprises the reference number of animals for reckoning the average annual performance of all cows.

Number of Dairy Cow Livestock Units (Dclu): This is the number of dairy cows present, expressed as "livestock units".

The large animal unit (LU) is a reference animal with an energy requirement of 3 000 fodder units (FU). It stands for a 600 kg dairy cow present on the farm for twelve months and producing 3 000 litres of milk with a fat content of 40 g, and one calf.

In using this unit, the number of dairy cows present is replaced by an equal number of standard dairy cows requiring 3 000 fodder units. The following formula is used to obtain the value of the LU equivalence ratio for a dairy cow:

$$K = 0.067\ W + (0.0002\ b + 0.005)\ Y + 0.21$$

in which:

K = equivalence ratio

W = weight in hundreds of kg

b = butterfat rate in grams per litre

Y = annual milk output in hl

In actual fact, the LU equivalences are approximative. Often a simplified table showing the data listed on page is sufficient.

Number of Bovine Livestock Units: This is the number of dairy cow livestock units plus other bovines for livestock or beef production, which are in turn expressed in LU, i.e. in cattle consuming 3 000 fodder units.

Of course the concept of LU based on the comparison of animals of different categories to a multiple or fraction of a cow with specific characteristics is debatable. Moreover, the energy requirements of the animal are simply a common denominator used for comparison. No account is taken of the fact that requirements may be satisfied by different types of fodder.

In any case, this system of equivalence does have the merit of providing an order of magnitude of herd size based on a single unit of measurement. This makes it easier to compare performances and numbers of units at the different work stations, and to calculate simple ratios involving several animal categories.

Relative Importance of Dairy Cows Within the Herd as a Whole: Dclu/Blu (%): This is the proportion of all bovine livestock units (BLU) represented by dairy cow livestock units (DCLU). This ratio illustrates the extent of dairy specialization.

A high ratio may indicate that reproduction is partly or just barely assured. A low ratio expresses the presence of a complementary livestock production activity such as young animals or fattening stock. Then again it may be an indication that the herd is beginning to grow.

Age at First Calving: This is the average age of cows giving milk for the first timme during the last fiscal year.

Average Lactation Rank: This is the sum R of lactation rankings in terms of the number of dairy cows present. This ratio indicates the average age of the dairy production unit, which is the outcome of renewal and culling decisions taken in the past.

An exact calculation would need to take into account how long each cow had held its present rank as of the date of estimation of average rank. In practice, one can use (R-0.5).

Herd Trends

The latest trends are shown by renewal and culling rates calculated over the course of the last fiscal year:

Renewal Rate: The renewal rate is the ratio of new milkers (cows giving milk for the first time) to the total number of cows in the herd.

The datum "age at first calving" may be added to give a fuller picture.

Culling Rate: This is the number of cows culled from the total herd of dairy cows during the past fiscal year.

Late culling of cows who calve regularly and on schedule and are consistently good milkers up to an advanced lactation rank, may be economically attractive. As long as the depreciation of the animal's value (annual drop in sale price when culled) is less than the extra gain from continuing to use the cow, there is no point in culling unless a breeding programme introduces high-potential breeding animals which promise to show profit in current variable charges.

Herd Reproduction

Technical control of breeding consists of measuring the rate of success at the different stages of fertilization, gestation and calving.

Fertility Rate or Successful Insemination And Gestation Rate: This is represented by the number of females calving with respect to the number of females inseminated. A high fertility rate depends on close surveillance of the cows, timely interventions (40 to 70 days after calving) when the animal is on heat, and satisfactory herd health.

Rate of Prolificity or Successful Calving: This is the proportion of viable newborn calves out of the total number of cows giving birth.

Fecundity Rate or Rate of Successful Reproduction: This is the proportion of viable newborn calves out of the total number of females mounted. It equals the sum of the rates of fertility and prolificity.

Average Interval Between Calvings for the Herd as a Whole (expressed in days): The target should be some 365 days, if necessary culling the least fertile females. Reducing the average interval between calvings by 18 days in a herd of 60 cows is no minor achievement considering that this means an additional 2 to 3 calves plus the extra milk produced.

X = average interval between calvings

Y = number of dry days (average)

Outside a range of:

360 days < x < 380 days and 60 days < y < 90 days, herd performance is not satisfactory, as this includes periods of underexploitation (or, exceptionally, over-exploitation) which lower the overall milk performance. The answer is animal hygiene, feeding, systematic fertility planning records, and also culling specific individuals for whom maintenance and veterinary costs are abnormally high for the milk they produce between calvings.

Estimation of Herd Worth

The herd represents a major part of the variable capital of the dairy farm. This capital should therefore be estimated at the start and close of the fiscal year.

Generally speaking, the livestock population of an agricultural enterprise is considered a stock subject to re-evaluation each financial year, rather than a capital asset.

There are two categories of livestock on the farm:

a. Breeding stock. These animals are present on the farm for several financial years. The value of the animal for inventory purposes will be taken as value at the time it enters the breeder category. This figure will remain constant until the animal is sold or dies.

b. (b) The replacement herd and dairy cows. The net sale value of each animal will be the figure retained for inventory purposes, with reference to market quotations on the day of the inventory. This figure is, of course, an estimate.

If the herd estimate at the end of the fiscal year is higher than at the beginning one will have:

Difference + Proceeds from animal sales - Value of animal purchases = Profit or loss from herd

Example:

Estimated value of herd at beginning of financial year	100 000
Estimated value of herd at close of financial year	130 000
Proceeds from animal sales	10 000
Value of animal purchases	50 000

30 000 + 10 000-50 000 = - 10 000

A loss of 10 000 on the herd is thus recorded for the financial year. This loss usually has two main causes-either the animals die during the financial year or there is a substantial drop in market quotations.

If the herd estimate at the end of the financial year is lower than at the beginning we will have:

Proceeds from sale of animals - Value of purchases of animals - Variance from estimation = Profit or loss on herd

Example:

Estimated value of herd at beginning of financial year	200 000
Estimated value of herd at end of financial year	180 000
Proceeds from sales of animals	70 000
Total value of purchases of animals	10 000

70 000-10 000-20 000 = + 40 000

During the financial year the farm therefore realized a profit of 40 000 on the value of the herd.

For "milk production", sales proceeds will be taken to mean actual sales receipts from off-farm sales and also fictitious sums which represent internal sales between work stations. Likewise, the value of purchases corresponds to real expenditure plus fictitious expenditure from internal sales between work stations.

It is also recommended that week-old calves who are to remain on the farm be listed as an internal sale between the "milk production" work station and the "livestock production" work station in order to differentiate clearly between milk production and livestock production costs.

Profits from the herd, milk sales and the sale of fertilizer will count as proceeds of the dairy side of the farm.

Should there be herd losses, the proceeds corresponding to the dairy side will equal:

Note 1

The dairy herd produces manure which is used as fertilizer on the farm to improve soils on which the fodder to feed the dairy cows is grown. Alternatively, the fertilized crops may be sold off-farm. This is why it is a good idea to reckon the value of manure, counting it as a proceed for the dairy farm (the system of internal sales will reckon it as a charge for user sections).

Feeding Livestock

Once the stock has been defined, the cropland to feed the stock needs to be designated.

In cost accounting, all operating expenses for this are attributed to milk production.

Energy Value of Feeds and Milk Production

Lucrative milk production consists of getting the highest possible yield at the lowest possible cost per litre of milk compatible with good herd health.

A balanced diet is a prerequisite to high yield. The feed ration is all feed distributed to the animal over a 24-hour-period to ensure coverage of maintenance and production requirements-milk or meat. Both quality and quantity are important in feed rations. The question is which feeds and how much.

Each feed ration has its characteristic:

- Energy level.
- Protein, fat, carbohydrate, mineral and vitamin content.
- Bulk, i.e. the volume of the ration (which depends on the fibre content or roughage).
- Livestock watering conditions.

Energy requirements may be evaluated in fodder units (FU). The reference element "barley" is used, in which the nutritional value of 1 kg of barley corresponds to one fodder unit.

Other feeds are compared to this standard:

1 kg of barley or 1 FU (1.65 kg bran
(0.96 kg wheat
(1.00 kg soyabean
(1.6 to 2 kg hay
(6.6 kg grass
(6.5 to 10 kg fodder
(1.25 kg oats
(3.125 kg oat straw
(0.94 kg soyabean or groundnut cake
(7.14 kg grass silage
(5.26 kg maize silage

Really strict equivalents using the fodder unit are not really possible because different animal species use what they digest in different ways. Another difference is whether the food is used for maintenance, growth, milk production, labour or gestation, etc.

One fodder unit for a fattening pig, for example, is 1 650 kcal and for bovines 1 883 kcal at average production levels.

In some countries, the energy value is indicated directly in kilocalories rather than in FU.

Maintenance requirements represent some 0.75 Fu per 100 kg of live weight.

Supplementary rations in the following proportions depending on the age of the dairy cow will also be needed:

2 years ... 1.00 supplementary fodder unit
3 years ... 0.60 supplementary fodder units
4 years ... 0.30 supplementary fodder units
5 years ... 0.15 supplementary fodder units

The figures also needed to be corrected during the gestation period (growth of the foetus). It is a good idea to add, at the different periods of gestation:

4 months ... 0.2 FU
5 months ... 0.4 FU
6 months ... 0.7 FU
7 months ... 1.0 FU
8 months ... 1.5 FU
9 months ... 1.5 FU

At current estimates, production of one litre of milk increases energy requirements by roughly 0.4 FU. The feed/production ratio is the first thing to look at in breaking down the per litre cost of feed.

A dairy cow needs fodder for maintenance. This can be thought of as a fixed cost inherent in the existence of the herd.

Maintenance requirements depend on the livestock species and the age and physiological status of the individual animal. A growing animal requires more feed to form new tissue and to grow. A gravid animal requires more feed for the growth of the foetus. The main contribution of any feed in excess of these requirements will be towards milk production.

When total feed consumption is compared to milk production, it becomes obvious that a good milker requires considerably less feed per litre of milk than a poor one (Graph 1). Production costs therefore tend to drop as milk cow yields rise.

This statement should really be qualified by the following comments:

- Costs differ substantially according to whether the diet consists of fodder grown on the farm, fodder purchased off-farm, coarse fodder or concentrates. So feed costs will vary greatly depending on what can be grown on the dairy farm and what is available on the local market.

Moreover, imported feed may be very cheap, as when it is subsidized by a producer country or enjoys preferential tariffs, but this situation can change from one day to the next, and the dairy famers' budget should include a contingency plan.

- Concerning coarse fodder, fodder type, fertilization, farming and harvesting techniques, etc. make for a considerable disparity in per hectare productivity, and this is also reflected in costs.
- Graph 1 shows neither the full complexity nor all components of the diet. The extra energy and nitrogen inputs of high-level production also entail extra costs.
- The linear representation of production/consumption used in most feed tables is of only relative value short of a given volume of production, so far as the genetic potential of the animal and the livestock producer's expertise at getting the most from his stock are concerned.

Selecting the Ration

The dairy cow is a converter of feeds which are, in the main, and out of necessity, produced right on the farm. These considerable masses of coarse fodder, which make up the basic diet, can only be grown near the livestock production area, and generally cost less per unit than commercial feed.

Grasslands have always been prime, potential, fodder-growing areas for dairy production. Dairy cow production is a natural choice in areas where soil and climate allow grass cover to thrive year round-or at least long enough to ensure partial coverage of fodder requirements.

Fodder conservation techniques are of considerable importance in producing surpluses to be kept and distributed at intervals over considerable periods of time. They allow a more constant livestock diet throughout the year, which is decisive in improving dairy performance.

Purchases of concentrated feeds or internal transfers of feeds produced on the multi-cropping livestock-production complex may be necessary to ensure balanced rations, because of animal intake capacity, fodder quality or carry-overs of fodder stocks.

Indeed, intensified dairy farming with supplementary feeding is only acceptable where economically justified, i.e. where the extra expense is offset by thc additional income.

The most precise economic and technical analyses are those covering individual animals over time at various stages in the milk and fodder production process.

In current practice, the livestock producer works out a type of ration for the different periods of the year in accordance with his stock's milking capacity and fodder supplies available on the farm. He estimates the fodder's nutritional value, i.e. the ability to cover a given period of milk production, and then supplements it by concentrates. An economic analysis of the ration should touch on both the suitability of the fodder system selected and the performance of the supplmentary feed. In economic terms this means the profitability of the various fodder/concentrates associations.

The technical objective is to provide the animal at all times with large amounts of farm-produced fodder. A sound fodder system is one which assures year-round, regular supplies, whilst matching the patterns and possibilities of thc farm.

There are no simple economic criteria for shared fodder systems. Production costs and the nutritional characteristics of the fodders within the system must both be borne in mind, plus the cost of supplementary feeding. A fodder budget based on feed rations plus information on expenses specific to fodder production can serve as the basis where what is involved is a single area on which different fodder combinations are grown. Indeed fodder budgets or operation models, based on technical and economic data, can suggest fodder systems appropriate for a specific farm and dairy establishment.

Going for intensive dairy production systems in most cases means setting aside a permanent grasslands area as the basic source of coarse fodder, at the same time improving grasslands productivity and adding new fodder crops. Even at highly intensive levels of dairy production, improved milk performance not only requires feed supplements in the form of concentrates, but also increased fodder production.

The Area under Fodder

The area under fodder consists of all cropland on the farm used for growing fodder.

The area under fodder includes:

1. The main fodder area (MFA), used primarily for growing coarse fodder:
    - Permanent grasslands (PGL) (pastures, grass, meadows, hay meadows, etc.).
    - Temporary and man-made grasslands.
    - Annual fodder crops (maize, fodder, grain silage).
    - Weeds used as fodder. The areas under grain for animals, when the grain is harvested dry, do not form part of the MFA.
2. The additional fodder area (AFA):
    - Intercropped fodder crops (two or more crops are grown on the same plot the same year, each of which is harvested at the same season. The crop with the longer growing cycle or the more important one is called the main crop and the others catch-crops.
    - The area under crops whose by-products are used for fodder: beet tops and pulp, pea pods, and so forth.

These two categories of fodder area taken together are often referred to as the total fodder area (TFA).

TFA = MFA + AFA

Fodder area costs are charged to livestock activities connected with the fodder area, the cost is proportional to the number of LU.

The farmer therefore needs to monitor the following:

- Milk production response to fodder intake.
- Performance of areas used for fodder production.

The usual technical yardstick is the average main fodder area/ livestock unit. This is the ratio between the number of dairy cow equivalents and the main fodder area.

Available on-farm Labour

In the developing countries, milk production tends to be extensive and so the goal of dairy development programmes is to improve productivity, thus reducing the labour component of production costs. One result of improved labour productivity is to expand the role of working capital.

The dairy farm takes on a new dimension and structures change. The same or a smaller workforce, the same or less land, much more capital, much more milk-all of which, in terms of production costs, modifies the pattern of expenditure. Service and equipment charges rise, depreciation costs and capital remuneration take on new weight, gains per unit produced are smaller but production volume shoots up. To sum up, earned income, which is what remains when all factors other than labour have been remunerated, is only assured or improved by boosting production and putting more capital into the new dairy undertaking.

In the developing countries, an active dairy policy entails:

- a sharp reduction in the number of workers for a given output;
- a reduction in the cost price of milk;
- greater labour efficiency.

This turnabout can only take place if the farmer has either his own capital or easy access to credit. So if a government wishes to develop milk production, it should make credit accessible to milk producers on preferential terms.

On small farms, the farmer and his family are the major source of labour and the corresponding costs are included in the auxiliary charges. On large farms, on the other hand, labour is distinct from capital. Workers receive a salary, but do not share in the proceeds of the enterprise.

For a comparison of the performance of highly different production units, therefore, all auxiliary charges will need to be considered and a reference unit chosen to identify jobs done on the farm. Reference units will vary from country to country in accordance with custom or tradition and, more frequently, in accordance with specific national legislation.

The term used in France is human labour unit (HLU) or labour unit (LU).

The human labour unit (HLU) is reckoned as equal to a total annual workload of 2 400 hours. The number of hours is subject to a corrective factor based on age and type of work.

Table 1: HLU for a full-time, able-bodied worker

Age		***Man or woman***
14 to 16 years	......	0.5
16 to 18 years	......	0.7
18 to 60 years	......	1.0
60 to 65 years	......	0.8-0.5
Over 65	......	0.5-0.3

Part-time workers are counted in fractions of LU based on the actual amount of hours worked on the farm.

The total annual workforce on the farm is a known factor when the reference unit is selected. Jobs should then be classified according to whether they are to be attributed to an activity or work section with or without distribution keys. These will be called either direct or indirect jobs.

This job classification assumes that a record is kept of job time, i.e. the time spent on a job directly attributable to a specific operation.

The number of HLU available on a farm is one of its definable criteria. The number of HLU may or may not be a limiting factor in a country or region. This depends on whether there is a possibility of hiring off-farm labour or not. Another factor is whether labour costs allow a profitable cost price.

Auxiliary Cost Centres

Auxiliary cost centres vary in number in accordance with the type of farm.

There is almost always a cost centre for farm equipment not specific to a given activity. This never includes a milking machine (used in only one work section), for example, but does include a tractor.

Here are other possible auxiliary cost centres:

- Water extracted by pumping from a well or reservoir, perhaps including water purification facilities.
- Site development
- Transformer
- Steam.

The costs of these auxiliary sections are then subsequently broken down by the various user sections. This paper covers only the agricultural equipment centre.

Agricultural Equipment Centre

A wide range of equipment including tractors, ploughs, cultivators, harrows, hoes, rollers, and fertilizer spreaders may be included in this section.

For the most important equipment such as the tractor, total costs must be related to production time and user sections subsequently billed pro rata by user hours.

However, if the farm were to plan to rent a tractor and driver from a third party for services rendered, the preceding hourly cost should also be billed as part of structural costs.

Structural Charges

The basic characteristic of structural charges (which primarily concern the structure of the farm) is that they are stable from one year to the next. In practice, however, this is a short-term stability, as any major decision by the farmer will modify the situation.

These charges chiefly include:

- Rentals and rental charges or capital remuneration;
- Floating labour;
- Land taxes and taxes not levied on a specific activity;

- Depreciation of administrative buildings and staff premises, depreciation on administrative equipment and miscellaneous;
- Various management expenses: transport, telephone, stamps, stationery, bookkeeping;
- Insurance;
- Financial costs;
- Costs which cannot be attributed to a specific activity (car or multi-purpose pickup, water, electricity, etc.).

Cost of Refrigeration, Milk Collection and Reception

Different Milk-collecting Systems

We shall first cover the main milk collection systems used by dairies and then try to break down costs in each in an attempt to help the industrial dairy farmer to choose among the various possibilities. In the simplest, door-to-door milk collection system, the milk truck stops at each farm, loads the milk cans and leaves an equal number of cleaned, sterilized empties.

This is a low-yield pick-up method, slow and interrupted by numerous stops. The transport vehicles wear out fairly quickly; the route may include bad roads. Also, the full carrying capacity of the trucks is not used because the milk cans, which come from different farms, are not all filled up to the same level.

Milk collection under such conditions is inevitably costly. But though it may be irrational, it is the only feasible method in many areas where a very large number of very small dairy farms are located in remote areas throughout the countryside.

Pick-up of uncooled or barely cooled milk in cans requires strict adherence to rules governing the duration of the collection period and the distance covered. Failure to observe these rules can easily alter the quality of the milk delivered to the dairy.

The method described is no longer valid in areas where people live close together or where small farms are grouped around villages. The best system in such cases is a collection centre to which farmers deliver milk every day, or preferably morning and evening, using their own means of transport. This system facilitates pick-up enormously as the milk truck loads the entire output of a village or milk-producing-area at a single stop. The time saved means that greater distances can be covered.

The structure of the collection centre may vary greatly depending on whether the milk is shipped from the centre to the dairy in bulk or in cans.

a. The milk may remain in the cans of individual farmers and the centre limit its role to providing a central pick-up point and cooling facilities, with a milk cooler its only working equipment. The milk delivered by individual producers will continue to be accounted for at the receiving platform of the dairy plant and the cans will be washed and sterilized in the plant. The disadvantage is that the milk truck loads a larger number of cans, some of which will not be filled to the brim.

b. The milk may be mixed at the collection centre. In this case the centre is responsible for keeping the accounts with individual milk producers. The centre may also wash and sterilize cans not to be returned to the plant. The centre may sometimes transfer the milk from the initial cans into larger-capacity pick-up cans. This type of centre is not advisable for handling amounts exceeding 1 000 litres/day.

c. For centres handling larger quantities of milk, the slow and fairly laborious practice of keeping milk in cans gives way to bulk holding.

In this case the centre's equipment will consist of a milk reception and weighing station; a milk cooler to lower milk temperature to some +4°C; insulated or refrigerated storage tanks for cooled milk; pumps and piping to distribute and transfer milk and a station with hot water, detergents and disinfectants for cleaning apparatus and milk producers' milk cans. Bulk conservation of cooled milk by the producer and transport in tank truck to dairy plant.

This system has the following advantages:

- Milking can be organized separately from milk collection, thus freeing the producer from delivery timetable constraints.
- The milk is cooled rapidly after milking which enhances its bacteriological quality.
- The longer-keeping stored milk can withstand a longer collection period which means the pick-up and delivery round can also be lengthened. Pick-up can be made every two days where milk quality permits (though a three day pick-up is definitely to be discouraged as the risk of psychrotrophic flora is too high).

- Milk reception at the dairy plant under this collection system is also different. Mainly in that the timetable for milk deliveries is more flexible, allowing more intensive use of dairy equipment, reducing the work force and increasing productivity all round. In technical terms, however, bulk collection is conceivable only under the following circumstances:
- The milk producers must have enough business to justify purchasing or otherwise making available a cooling tank. This also presumes a rural power network sufficiently developed and reliable to accommodate the power demand from cooling tanks without excessive risk of blackouts.
- The milk in each tank must be of good quality to avoid any risk of contaminating the milk in the tank truck.
- The tank truck must be able to back right up to the cooling tank (maximum distance 8–15 m).

In areas accessible only to small trucks working with milk cans, bulk pick-up with large-capacity, heavyweight tank trucks is simply not feasible.

It should be remembered that all-zone bulk pick-up will rarely be possible. The first consideration in choosing a milk collection system should be to maximize both technical and economic advantages. Where possible, the collection area selected should lend itself to gradual expansion.

In a uniform collection zone, the dairy plant can set up a single collection system. Unfortunately, in the Third World the dairy industry is frequently part of a development process which includes milk producers of all sizes and so the plant will have to adopt a mix of approaches:

- In a specific collection zone where large milk producers are in the majority, the producers are equipped with cooling tanks, and pick-up is by tank trucks. Tank trucks may have an additional rack for milk cans from small producers. Barring that, the milk produced by several farmers will have to be picked up at a central collection point. A combination milk can/refrigerated tank truck pick-up is risky because of possible contamination and overheating.
- Where small producers are in the majority, the plant may have tank trucks to collect milk from several collection centres as

well as from one or more large producers equipped with cooling tanks.

As they work out their refrigeration and collection costs, some dairies may even have to contemplate the following option for areas where access is difficult:

a. Setting up a collection centre
b. Small-scale or cottage industry cheese manufacture at milk production sites as a means of reducing transport costs.

Making the most of Milk Pick-up Rounds

Whichever pick-up system is chosen, the cost of milk collection is clearly almost entirely dependant on:

- the truck (depreciation, maintenance);
- labour.

The goal of sound management should therefore be to minimize these two major cost items. The first thing to do is to optimize milk pick-up rounds and then:

- identify the number and size of trucks needed;
- minimize the working hours of the milk collector whilst assuring the delivery of high-quality milk.

To optimize milk pick-up rounds, the milk collection area must be mapped out. This involves:

- Classifying collection sites by village, hamlet, isolated farm or collection centre.
- Classifying the number of milk producers at each site in accordance with maximum milk output per producer. Where necessary, milk quality may also be indicated.
- Totalling peak and slack period milk totals for each collection point.
- Listing the peak load of the various types of trucks or tank trucks available on the market.

With these elements in hand, several milk round models can be "tested". At this point, there is a temptation to reduce rounds to the minimum number by lengthening their duration, using bigger trucks and having longer intervals between pick-ups.

At this stage, economic factors cease to be the sole consideration. Depending on milk quality and the collection system selected, the maximum interval between two pick-ups (daily or every two days) is the next thing to be worked out, remembering that pick-up frequency has a major bearing on costs.

In fact, in two-day pick-up service, more milk is collected, necessitating bigger trucks and closer pick-up points.

Any attempt to maximize the length of milk rounds is limited to how long the milk can be kept without cooling it: this is particularly true of milk cans. This interval depends on how big and fast the truck is and on milk production density within the pick-up area. Maximum round length may also be limited by the following factors:

- No round can exceed the maximum number of hours the driver is legally allowed to work each day.
- The plant may be organized in such a way that milk rounds need to end at specific times during the day.
- Timetables of access to farms.
- Highways or roads which can only be used during the day, or at certain times of the day.

The above limiting factors on the length of milk rounds need also be considered in determining truck size.

One practical way of deciding which of the different round patterns "tested" is more attractive, not forgetting the above imperatives, is to compare them as follows:

The total mileage for each round is divided into the active sector, i.e. the mileage between the first and last milk producer (kr) and the downtime zone, meaning the distance between the dairy and the first producer and between the last producer and the dairy (kt).

The number of litres of milk picked up daily per kilometre of active distance

$$\frac{\text{total litres}}{\text{kr}}$$

Corresponds to what is often designated as production density or technical density.

The number of litres of milk picked up daily per kilometre of active distance plus downtime zone:

$$\frac{\text{total litres}}{\text{kr+kt}}$$

Which constitutes the pick-up density.

Round optimalization makes it possible to increase pick-up density without increasing vehicle size. The direction of the pick-up circuit should permit the maximum average truck speed and the shortest standing time for milk.

Two plants who whish to optimize their milk rounds should avoid overlapping and superimposed pick-up zones.

Rounds have been worked out "manually", and often very cleverly by professionals since the beginning, but today the use of information technologies and sceintific estimates can do a quick, flexible, highly effective job.

For correct optimization of pick-up rounds, the various segments of pick-up working time require consideration.

Pick-up Work-time Study

In optimizing pick-up rounds, the total time necessary for the round should be taken into account. This interval can be divided into:

- The collector's time in the truck.
- The collector's time at farms or collection centres.

Time spent in the truck can, in theory, be calculated by measuring distance, truck speed and road conditions. Another simple method is to use a chronometer.

Time spent at the farm or collection centre should be split into major operations for greater precision. These major operations will differ in accordance with the collection system selected by the plant.

For bulk collection the major operations may be divided into:

- Preparation
- Pouring out the milk
- Final operations
- Sampling
- Sales to producers
- Breaks, downtime.

The first three are compulsory, the last three optional and there may be a certain overlapping between some operations.

The three compulsory operations are classified into two types: those where the time interval is proportional to the amount of milk (pouring out the milk) and those where the time interval has nothing to do with the amount of milk. The optional operations are fixed-time operations. Fixed-time intervals are far more important for tank truck pick-ups than variable intervals depending on the amount of milk delivered. It is essential to divide the time into fixed-times and variable intervals proportionate to the amounts of milk collected. This is so that time can be budgeted to account for eventual changes in the amounts of milk collected, and so as to predict the length of time necessary to pick up different hypothetical amounts of milk.

Fixed can-pick-up times are relatively inflexible as they do not include truck start-up and stop time. If the pick-up is organized by the dairy plant, milk samples are always taken at the plant.

Optional operations are included fixed times and, as for bulk collection, include:

- Sales to producers
- Breaks, downtime.

Fixed Times for Bulk Collection

Fixed times include preparation:

- Getting out of the truck
- Unrolling the milk hose
- Inserting the vacuum pump hose or coupling the pump intake valve.
- Reading the gauge (unless the truck has a counter) and recording milk intake.
- Starting the filling operation (opening the vacuum pipe or switching on the pump).
- At the start of the pumping operation, taking a milk sample, putting it in a flask and putting the flask away (optimal).

A study carried out in France showed that the overall preparation time per producer ranged from 0.2 to 4.2 minutes-the average time was 0.96 minutes.

The most important variable factors are:

- The distance from the milk holding tank to the tank truck parking site.
- Accessibility of the tank
- The location, quality and number of attachments of the tank truck and tank.

- The collector's skill and ability to organize operations.

Variable factors also include the length of the final operations:

- Turning off the pump (where there is a pump) or closing the vacuum intake valve.
- Possible quick rinse of the tank or turning on the programmer.
- Rolling up the hose and vacuum pipe.
- Reading the counter, if there is one, if it lacks an automatic recording device; where applicable, return of ticket to producer or recording the amount of milk on the collector's sheet.
- Getting back into the truck.

A French study showed that the time spent on these final operations ranged from 0.15 to 3 minutes, averaging 0.89 minutes. Where the tank was rinsed out, from 0.7 to 0.8 minutes was required for this operation alone, nearly doubling the total final operation time.

Taking milk samples is an optional fixed-time operation. A study produced the observation that hand-sampling required from 0.07 to 2.66 minutes, averaging some 0.78 minutes. Obviously, automatic or semi-automatic samplings are a real time-saver.

Proportional Intervals for Bulk Collection

Milk pumping time, the only time period which varies with the amount of milk delivered, also varies with the capacity of the pump or vacuum system and with hose length and diameter. Where these factors are constant, the interval varies only with the amount of milk pumped, though in practice, it is highly variable.

As a basis of estimation, a reasonable interval would be some 0.60 – 0.85 cm/m/1 of milk.

Services and Supplies

Where the milk is collected in cans, the time spent returning the cans is always extra, whereas in bulk collection return time and milk pumping time can overlap.

How many products are distributed, and whether the collector is paid or just keeps a record of merchandise sold has a great bearing on the time taken by this operation.

Can/return time, however, needs to be accounted for separately so that the cost can be compared with the profit from other services and supplies. (These may be dairy products produced by the plant, detergents for cleaning the tank and milking machine, brushes, etc.)

Well-organized supplies are a source of profit for the dairy plant and are recognized as a service rendered to milk producers. A separate management system should be organized for such services.

Breaks, Downtime

It is true that downtime must be kept to a minimum. Quite a number of dairy plants, however, noticed that the collector was the only person at the plant who had any real contact with the milk producer, and this gave them the idea of using the collector as an extension agent. Indeed, who better than the collector is in a position to spend a few minutes to pass on some message on milking hygiene, the use of concentrates for dairy cows, and so forth. Spectacular results have been obtained in achieving specific targets set by the plants for milk collectors.

The time spent by the collector in "training and/or informing" producers should be accounted for under some such heading as "Training milk producers".

Note: The need for optimizing milk rounds and job times needs to be explained to collectors. They must abide by the schedules of when rounds begin and when the milk is delivered to the plant. They also need to be able to justify any delay or time overrun within the management audit context.

Analysis of Collection Costs

The various items under collection costs depend largely on whether milk is collected in cans or in bulk; the following considerations will help identify the corresponding costs:

- Tank truck pick-up is often one way of avoiding Sunday collections but it needs to be remembered that the Saturday and Monday pick-ups will be 50 percent bigger than on the other days of the week.
- Before opting for bulk pick-up, thereby completely modifying the equipment and methods of producers, collectors and plant, it would be good to get as much information as possible on the financial consequences of bulk operations. Usually, the change over to a bulk collection system comes about when the equipment of the older pick-up system has either depreciated or can be switched to another pick-up zone. A shift to bulk collection in a given area should be preceded by a global study, even where bulk collection is only to be sectional.

- The term peak or seasonal imbalance coefficient is given to the ratio:

$$W = \frac{Q_1}{Q}$$

Q_1 corresponds to the milk supply in litres of the day or days of heaviest pick-up.

Q_2 corresponds to the yearly pick-up of litres of milk divided by 365.

The fleet of vehicles can be designed to allow peak-day pick-up rounds to coincide with those of slack production periods. Vehicles are only used at maximum capacity Q_1 during the peak period. On the other hand, it may be attractive in some instances to assume that the milk trucks will make two rounds at nominal load Q_1 during peak days, and plan the vehicle fleet accordingly. In this case the number of trucks necessary will be:

$$\frac{Q_1}{2\,Q_1}$$

Truck or tank trunk capacity, and thus the number of vehicles needed, has a great bearing on collection costs. It is therefore advisable to look for a model which can minimize collection costs per hectolitre and which bases the analysis of cost forecasts for milk collection on anticipated volumes. This involves "testing" the various possible types of vehicles. Usually, however, volume is not an exact known quantity so there is an assumed floor and an assumed ceiling. Three such tables are therefore necessary to verify the low-, medium-and high-volume assumptions.

Labour costs are considered fixed costs in some plants and variable costs in others. Indeed, where collectors work only a few hours a day collecting milk and spend the remaining time distributing dairy products or concentrated feeds, milk collection time can be considered proportional to the amount of milk collected, and labour costs should be listed under variable costs. If, however, the collector's free time is used for minor maintenance of and cleaning his truck, labour should be considered a fixed cost.

Variable costs for trucks should be reckoned so as to show the per kilometre cost of each truck for each litre of milk collected.

A portion of the item "structural costs" needs to be included under fixed costs. The following structural costs will be invoiced to milk collection as fixed costs:

Management of the "Milk Production" Department

This service does in fact have its own staff for managing the fleet of vehicles and service staff, and handling relations with producers. This service will have its own premises, and such various management costs as:

- General management
- Purchasing service (vehicles, cans, ...)
- Personnel service.

Table 2: Collection Cost (milk pick-up in cans)

Description	*Amount*
Fixed Costs	
Depreciation:	
-Trucks	......
-Cans	......
-Garages or parking areas	......
-Tools	......
Insurance and taxes: trucks	......
Garage and parking costs:	......
-Maintenance	......
-Electricity	......
-Heating (where necessary)	......
Labour	......
Invoicing of part of structural costs	......
TOTAL FIXED COSTS	......
VARIABLE COSTS	
Fuel	......
Oil	......
Tyres	......
Vehicle maintenance and repair costs	......
(spare parts and labour)	......
Can maintenance costs	......
Tool maintenance costs	......
TOTAL VARIABLE COSTS	......
TOTAL COSTS (FIXED + VARIABLE)	......

Analysis of Cooling Costs

The Various Cooling Systems

Farm cooling systems use mechanical compression. The basic principle of these machines is to lower temperatures through forced evacuation of an appropriate liquid refrigerant, usually freon. All compression equipment includes the following:

- An evaporator, usually a coil, in which the liquid refrigerant is evaporated for quick cooling of the ambient temperature.
- A compressor which acts as a suction and force pump, taking in the steam from the evaporator and compressing it in the condensor.
- A condensor or liquefier externally cooled by water or air in which the fluid is liquified under pressure.
- A trigger valve regulating the return flow of the fluid in the evaporator where it is reconverted into steam.

In direct expansion refrigeration, the most commonly used type, the cooling medium, i.e., milk, is in direct contact with the evaporator.

In other cooling machines, some other liquid, which may be water, is cooled by direct expansion. This is then used to cool the milk.

In bulk collection, the milk is kept in direct expansion cooling tanks. If, on the other hand, the milk is collected in cans, the cans are set in a milk cooling vat with a flow of chilled water. Just for the record, there is also the heat plunger (can immersion coolers), which is a direct expansion apparatus with a condensing unit and a rod attached by a flexible tube. This rod includes an evaporator and a shaker. The heat plunger may be introduced directly into the can but this practice is not recommended for bacteriological reasons. The thermal plunger system can be used to cool the water in the can immersion tank.

Bulk Collection: Selection of Cooling Tank

The first thing is to select the volume of the cooling tank. Remember, cooling and pick-up are two parts of a single system. The cost of the alternate day tank is higher then that of the every day tank, but the system does help minimize collection costs.

Tank volume depends on the producer's needs. The coefficient of use (U C) of each tank must be calculated.

The following points are important:

- Collection is a money saver where large amounts of milk are collected at each pick-up point.

Apparently, some 150 1 of milk per producer each day is the minimum figure at which bulk pick-up represents savings.

- Bulk collection is a money saver, in absolute terms, under less favourable collection conditions, where many miles need to be covered. Less money is saved, however, when pick-up is slowed by the need to serve a great many small producers.

Expenditure of Energy for Milk Cooling

Trials at the CTGREF in Anthony (France) on two cooling tanks with capacities ranging from 200 to 6 140 1 at + 20°C ambient temperature (and for cooled milk from 35 to 4°C) indicate:

a. An average consumption of 19 Wh/1 for alternate day utilisation.

b. This consumption figure drops as tank capacity increases. For every day use, the figure goes from 25 ha/1 for capacities under 300 1, to 15 W/h/1 for capacities over 3 000 1. For alternate day use, under the same conditions, the figure goes from 30 to 20 Wh/1.

c. This consumption fluctuates considerably around these average values with the quality of the equipment ranging from 10.8 to 30.9 ha/1 for every day use and from 14.7 to 36 Wh/1 for alternate day use, without these ranges being significant in terms of tank capacity or their cooling system (direct expansion or ice bank system).

In addition to providing users a wider selection of better-performing equipment the following are recommended ways of lowering consumption:

a. Never exceed alternate day storage. Collecting milk every three days is unacceptable not only on milk quality grounds: power consumption increases to the point where it is no longer offset by lower transport costs.

b. Develop energy recovery systems. Two complementary techniques can be used at the dairy plant for this purpose: chilled water pre-cooling of milk, and heating water with the calories recovered from the the condensing unit of the tank.

The objective is cheap removal of negative kilo-calories from the chilled water to cool the warm milk by a plate heat exchanger which lowers the temperature to under 20° in just seconds.

Some two to two-and-a-half times as much chilled water as milk is necessary. The milk is pumped intermittently through the heat exchanger by the milk pump from the extractor of the milking machine. The milk outlet of the exchanger requires a flow regulation valve to set up a slight counter pressure. The water outlet has an electromagnetic valve to avoid wasting chilled water. The exchanger requires a filter to avoid rapid clogging of the pre-cooler.

Fuel savings mount as the temperature of the cooling water drops, so the system is an unattractive solution in hot countries.

Heating Water from Calories Recovered from Condenser

As negative kilo-caleries are produced by the evaporator, calories are produced and evacuated by the condenser. This is, in fact, the principle of the heat pump.

From the calorie standpoint, the amount of heat given off by the condensor equals the amount of heat extracted from the cooled medium plus the energy absorbed by the compressor. Globally, one kilowatt supplied to the compressor is considered to allow recovery of from 1 to 2.5 kilowatts from the condenser.

The hot (80°C) gases from the compressor are directed to a tubular heat exchanger where they give up their calories to the chilled water, heating it. The gases then move across the air condenser to the tank evaporator where they expand.

Where an accumulator-type regenerator is used (Diagram 3), the exchanger is placed inside a heat accumulator tank. A well-chosen system can produce water which retains a constant temperature of 60°C until the next milking from each batch of milk cooled. If desired, even hotter water can be obtained (up to a maximum 80°C). During periods of low milk production, an electric coil can also be added to this system to top up the heat supply. Pre-cooling milk in chilled water is only appropriate for large herds or collection centres handling at least 2 000 1/day of milk. Heating water by means of heat recovery from the condensor, however, is a suitable technique for many dairy plants. Each individual country needs to do its own economic investigation of investment and power costs.

Refrigeration Cost Components

Refrigeration costs include fixed costs and variable costs:

a. Fixed costs

Fixed costs, which are independent of the amount of milk cooled, include:

- Depreciation of the tank (including interests on loans requested or remuneration of invested capital). The usual amortization period is five years.

In consideration of seasonal variations some milk producers buy large tanks, though this increases refrigeration costs. Financing the investment is always a problem. The burden may fall on the dairy, if it is lending the tank to the producer, or upon the producer who will then expect to recover his expenditure and be compensated for his efforts to improve milk quality.

- Depreciation on the building in which the tank is kept.
- Tank cleaning costs.

This item should be listed under fixed costs as cleaning is compulsory after each milk pick-up irregardless of the volume of milk produced between two collection periods.

Tank cleaning, whether manual or automatic, includes:

- labour
- cleaning agents
- cold water
- cost of heating water
- power consumption of pump for automatic washing.

b. Variable costs

Variable costs depend on the amount of milk cooled. They include:

- Power consumption
- Maintenance costs.

This is an undefined item which depends on the specific operating conditions of each apparatus.

Cooling equipment is so highly specific and milk quality so dependent on a smooth-running cooling system that a great many dairies prefer to take on the responsibility for maintenance of the cooling tanks of their suppliers. This service may be provided free of charge. On the other hand it may be invoiced, in which case there are two possible systems:

- invoicing separate interventions at real costs;
- a maintenance contract.

In the latter case, the producer pays a set sum each year and there is no separate charge for servicing.

In this case "maintenance-repairs" is classed under fixed charges.

Sharing Refrigeration Costs between Producer and Plant

Either the dairy plant or the producer may be responsible for financing the cooling tank.

The producer purchases the cooling tank himself. In most cases, the dairy pays a compensatory sum for each litre of milk produced. The producer should deduct any compensatory sums paid by the dairy when reckoning refrigeration costs.

It must also be remembered, particularly where bulk pick-ups are concerned, that milk is almost always priced in accordance with quality. One common criterion is the total microflora content in the milk. Refrigeration helps to retard bacterial development, thus ensuring that the producer gets a better price for his milk.

The producer should consider per litre price differences for different grades of milk in calculating real costs for refrigeration.

The investment cost is borne by the dairy.

In this case, the dairy opts for one of the following three approaches:

- The tank is lent to the producer free of charge.

The drawback for the dairy is that it is the sole investor. The advantage is that it can tailor tank capacity to the variable production of individual producers and adapt to changes in the collection system, i.e., modification of the milk rounds or different pick-up schedules.

The dairy does not have to pay compensatory sums to producers in this case.

- Leasing the tank to the producer.

Leasing may take the form of a fixed monthly payment or a discount on each litre of milk produced.

As in the preceding instance, the dairy remains the owner of the tank.

- Leasing/sale of the tank.

This follows the same pattern as the other two cases but after two, three or five years, the tank becomes the property of the milk producer.

Overall cost "Refrigeration-Collection-Reception"

The cost of milk can be reckoned at the farm or at the dairy plant reception platform.

Our study shows that proper organization of "refrigeration-collection-reception" is quite complex. It seems that "collection-reception" should definitely be the responsibility of the dairy. It would therefore be advisable to determine milk costs at the farm level, where possible. In this way, even if independent truckers are responsible for pick-up, transport operations will be under the authority and supervision of the dairy.

The global "refrigeration-collection-reception" cost will be calculated by the dairy which will retain out of refrigeration costs only the share for which it is responsible. Global cost is the item which allows the dairy to compare the merits of different collection systems.

It needs to be stressed that a well-organized cooling and collection system has a major bearing on milk quality and hence on processing costs and the quality of finished products. Therefore not only costs, but also the higher-quality finished product, which is not always easy to express numerically, need to be considered.

In bulk collection, there are slight differences between the amounts of milk for which the producer is paid and the amounts actually reaching the plant. These milk losses need to be calculated for each round. If the dairy organization has several plants, quantity control should make it possible to compare plant performance.

The Milking Machine and Milking

The milking process is the collection of tasks specifically devoted to extracting milkfrom an animal (rather than the broader field of dairy animal husbandry). This process may be broken down into several sub-tasks: collecting animals before milking, routing animals into the parlour, inspection and cleaning of teats, attachment of milking equipment to teats, extraction of milk, removal of milking equipment, routing of animals out of the parlour.

Milking Schedules

Maintaining milk yield during the lactation period (approximately 300 days) requires consistent milking intervals, usually twice daily and with maximum time spacing between milkings. In fact all activities must be scheduled around the milking process on the dairy farm. Such a milking routine imposes restrictions on time management and personal life of an individual farmer, as the farmer is committed to milking in the early morning and in the evening for seven days a week regardless of personal health, family responsibilities or social schedule. This time restriction is exacerbated for lone farmers and farm families if extra labour cannot easily or economically be obtained, and is a factor in the decline in small-scale dairy farming. Techniques such as once-a-day milking and voluntary milking have been investigated to reduce these time constraints.

Automation in Milking

To alleviate the labour involved in milking, much of the milking process has been automated: many farmers use semi-automatic or automatic cow traffic control (powered gates, etc.), the milking machine has entirely automated milk extraction, and automatic cluster removal is available to remove milking equipment after milking. Automatic teat spraying systems are available, however there is some debate over the cleaning effectiveness of these.

The final manual labour tasks remaining in the milking process were cleaning and inspection of teats and attachment of milking equipment (milking cups) to teats. Automatic cleaning and attachment of milking cups is a complex task, requiring accurate detection of teat position and a dextrous mechanical manipulator. These tasks have been automated successfully in the voluntary milking system.

Automatic Milking Systems (AMS)

Since the 1970s, much research effort has been expended in investigating methods to alleviate time management constraints in conventional dairy farming, culminating in the development of the automated voluntary milking system (AMS).

Voluntary milking allows the cow to decide its own milking time and interval, rather than being milked as part of a group at set milking times. AMS requires complete automation of the milking process, as the cow may elect to be milked at any time during a 24 hour period.

The milking unit comprises a milking machine, a teat position sensor, a robotic arm for automatic teat-cup application and removal, and a gate system for controlling cow traffic. The cows are permanently housed in a barn, and spend most of their time resting or feeding in the free-stall area.

When the cow elects to enter the milking unit (due to highly palatable feed that she finds in the milking box), a cow ID sensor reads an identification tag on the cow and passes the cow ID to the control system. If the cow has been milked too recently, the automatic gate system sends the cow out the unit. If the cow may be milked, automatic teat cleaning, milking cup application, milking, and teatdipping takes place. As an incentive to attend the milking unit, concentrated feedstuffs needs to be fed to the cow in the milking unit. The barn may be arranged such that access to the main feeding area can only be obtained by passing the milking unit. This lay out is referred to as forced cow traffic.

The innovative core of the AMS system is the robotic manipulator in the milking unit. This robotic arm automates the tasks of teat cleaning and milking attachment and removes the final elements of manual labour from the milking process. Careful design of the robot arm and associated sensors and controls allows robust unsupervised performance, such that the farmer is only required to attend the cows for condition inspection and when a cow has not attended for milking.

Typical capacity for a AMS is 50-70 cows per milking unit. AMS usually achieve milking frequencies between 2 and 3 times per day, so a single milking unit handling 60 cows and milking each cow 3 times per day has a capacity of 7.5 cows per hour. This low capacity is convenient for lower-cost design of the robot arm and associated control system, as a window of several minutes is available for each cow and high-speed operation is not required.

AMS units have been available commercially since the early 1990s, and have proved relatively successful in implementing the voluntary milking method. Many of the research and developments have taken place in the Netherlands. The most farms with AMS are located in the Netherlands and Denmark.

Advantages

Elimination of Labour: The farmer is freed from the milking process and associated rigid schedule, and labour is devoted to supervision of animals, feeding, etc.

Increased Milking Frequency

Milking frequency may increase to three times per day, however typically 2.5 times per day is achieved. This may result in less stress on the udder and increased comfort for the cow, as on average less milk is stored. Higher frequency milking increases milk yield per cow, however much of this increase is water rather than solids.

Perceived Lower Stress Environment

There is a perception that elective milking schedules reduce cow stress. A study found no decrease in stress between automatic and conventional milking.

Herd Management

The use of computer control allows greater scope for data collection. Such data allows the farmer to improve management through analysis of trends in the herd, for example response of milk production to changes in feedstuffs. Individual cow histories may also be examined, and alerts set to warn the farmer of unusual changes indicating illness or injury. Information gathering provides added value for AMS, however correct interpretation and use of such information is highly dependent on the skills of the user.

Considerations and Disadvantages

High initial cost -AMS systems cost approximately €120,000 ($190,524) per milking unit (presuming barn space is already available for loose-stall housing). Equipment costs decreased from $175,000 for the first stall to $158,000. Equipment costs decreased from $10,000/ stall for a double-six parlour to $9000/ stall for a double-ten parlour with a cost of $1200/ stall for pipeline milking. Initial parlour cost was increased $5000/ stall to represent a high cost parlour. Whether it is economically beneficial to invest in an AMS instead of a conventional milking parlour depends on constructions costs, investments in the milking system and costs of labour. Besides costs of labour, the availability of labour should also be taken into account. In general, an AMS is economically beneficial for smaller scale farms, and large dairies can usually operate more cheaply with a milking parlour.

Increased Complexity -While complexity of equipment is a necessary part of technological advancement, the increased complexity of the AMS milking unit over conventional systems, increases the reliance on manufacturer maintenance services and possibly increasing operating costs.

The farmer is exposed in the event of total system failure, relying on prompt response from the service provider. In practice AMS systems have proved robust and manufacturers provide good service networks. Because all milking cows have to visit the AMS voluntarily, the system requires a high quality of management. The system also involves a central place for the computer in the daily working routines.

Difficult to Apply in Pasture Systems -AMS works best in zero-grazing systems, in which the cow is housed indoors for most of the lactation period. Zero-grazing suits areas (e.g. the Netherlands) where land is at a premium, as maximum land can be devoted to feed production which is then collected by the farmer and brought to the animals in the barn. In pasture systems, cows graze in fields and are required to walk to the milking parlour. It has been found that cows tend not to attend the milking unit if the distance to walk is too great. There is currently a research project at the Dexcel facility in New Zealand where cattle are on pasture and milked by AMS.

Lower Milk Quality - Somatic cell count (SCC) and Plate loop count (PLC) are, respectively, measurements of the quantity of white blood cells and total number of bacteria present in a milk sample. A high SCC indicates reduced udder health (as the immune system fights some infection) and implies lower milk quality. AMS herds consistently show higher SCCs than conventionally milked herds. A high PLC indicates bacterial contamination, usually through poor sanitation or cooling and similarly implies low milk quality. High PLC in AMS may be attributed to the continuous use of milking lines (rather than twice a day in conventional systems), which reduces the time window for cleaning, and the incremental addition of milk to the bulk milk tank which may not cool efficiently at low milk levels.

Possible Increase in Stress for Some Cows -Cows are social animals, and it has been found that due to dominance of some cows, others will be forced to milk only at night. Such behaviour is inconsistent with the perception that AM reduces stress by allowing "free choice" of milking time.

Decreased Contact between Farmer and Herd -Effective animal husbandry requires that the farmer be fully aware of herd condition. In conventional milking, the cows are observed before milking equipment is attached, and ill or injured cows can be earmarked for attention. Automatic milking removes the farmer from such close contact with the animal, with the possibility that illness may go unnoticed for longer periods and both milk quality and cow welfare suffer. In practice, milk quality sensors at the milking unit attempt to detect changes in milk due to infection, and farmers inspect the herd frequently. However this concern has meant that farmers are still tied to a seven-day schedule.

Record Keeping

Farm records are kept for all or some of the following reasons:-

To satisfy the Receiver of Revenue: This is an essential requirement of record keeping but should not be the sole reason, and a record system can be designed which satisfies the Receiver and is also useful for other purposes.

To Assist in Financial Planning Decisions: Financial records, in more detail than those required for the Receiver, can be used for cash flow planning, enterprise analysis and other purposes.

To Control Labour: *This is usually a wages book recording days worked, wages paid, money owed, leave* etc.

To Assist in Land Management Decisions: These include farm maps and grazing, irrigation, fertilizer use, crop yield, areas and management operations records.

To Assist in Livestock Management Decisions: *These are the records of individual animals and groups of animals, their production, health, feed use* etc.

No Logical Reason: A lot of useless information is often kept which is never, or can never, be converted into useful information.

Criteria for Record Keeping

The brief summary of record types given above illustrates that several sets of records must be kept, inevitably involving much of the farmer's time. Hence, if records are not to be more trouble than they are worth, they should satisfy the following criteria:

They Must be Useful

Unless data which is being recorded will at some future time be used (turned into information) in making management decisions it should not be recorded at all.

Records Must be Kept in such a form that they can be easily Converted into Information: Before keeping a record, the eventual end use must be decided upon so that the form in which the data are recorded will facilitate later analysis and interpretation. Too often the end use is not considered, and the usefulness of the data is severely impaired.

Record Keeping Systems Must be Simple

Dairy farmers have enough to do without burdening themselves with complex record keeping systems, that are difficult to understand

and time consuming to complete, and therefore nearly impossible to delegate to employees.

Duplication Must be Avoided as much as Possible

Some data may have to be recorded more than once in different forms, but this must be reduced to a minimum. In other words, if a record is to be made in the field, the recording system should be such that data can be conveniently entered in the field and does not have to be re-entered back at the office.

Records must Lead to Actions being taken

Unless a record is specifically intended to be used for some future action or in management planning it should not be kept.

A Simple set of Dairy Herd Records

An Overview

It is not the purpose of this guide to describe a complete farm record keeping system. A simple, effective dairy herd record system, which satisfies the criteria listed above, will be described.

This convenient system can be used to supply information for:-

- Effective monitoring of animal performance right from birth.
- Evaluation of management and feeding systems.
- Individual animal comparisons to assist in breeding, culling and other decisions.
- Breed society, milk recording and computer program usage.
- Extraction of useful herd indices for evaluation and comparison.
- Production of Action Lists for management.

The physical requirements for the system are cow byre sheets (or milk recording book), individual animal cards, ring-binder file/s with plastic filing sleeves, cow calendar board, heifer-rearing board, an insemination book, daily diary and last, but not least, a comfortable office. Other records which may be added include a separate mastitis record and chalkboards for drying-off and feeding lists.

Cow Byre Sheets/Milk Recording Book

These are used for:

- Recording of live mass
- Recording of condition scores

- Recording of butterfat percentages
- Recording of daily milk yields
- Recording concentrate allocations
- Recording concentrate use
- Noting drying-off dates
- Noting oestrus
- Roughage record (camps grazed, quantity silage fed etc.)
- Mastitis presence and treatments
- Running lactation totals.

This layout can be achieved using some of the sheets which are given away by feed companies, or certain bookkeeping analysis books sold by commercial stationers can be used.

The essential points are as follows:-

- Cow names are recorded vertically in alpha-numeric order. Milk recording sample bottle numbers, butterfat percentages, live masses and condition scores are recorded monthly.
- Condition scores may be recorded fortnightly
- Days of the month are recorded horizontally, ideally with four columns per day. This is usually quite simple as many of the sheets which are available have spacc for three milkings per day plus a total and, as most herds are milked twice daily, the spare column can be used for two purposes.
- The far right hand side is used for running lactation totals. Most people find it far easier to add vertical columns of numbers than horizontal columns. The most frequently done additions are thus arranged vertically. The only horizontal addition that has to be done is the running lactation total. This is done because official Milk Recording results frequently come back too late to be used for culling decisions and for other analyses.

There are very few farms where the milk is recorded at every single milking. In any event adding every single day's yield for the purpose of estimating a cow's yield is unnecessarily laborious (Official Milk Recording only uses one day per month).

Therefore, it is suggested that three days of the month are selected; for example, the 3rd, 14th, and 24th. Add the three days together, and multiply by 10. If it's a thirty day month the total for the month has been arrived at, which can be added to the total for the previous month. If it is a thirty-one day month, add the three days together, multiply by 10 and add the middle day to get an estimate for the month. For February add the three days together, multiply by 10 and subtract two days yield. For cows which calved or were dried off during the month, add the available days, divide by the number of records used and multiply by the number of days the cow was in milk.

An example:-"Betty's" yields were (33.75l + 33.5l + 27.25l) x 10 = 945l for a 30 day month or, 945l + 33.5l = 978l if it were a 31 day month.

Another example:-"A59" calved on the 10th, her yields for the available days were 17 and 25. It was a 31 day month, she was therefore in milk for 21 days and her estimate would be 17l + 25l = 42 x 21 ¸ 2 = 441l.

It is useful, whenever possible, to add the vertical columns of yields as a check against milk sales or the inaccuracy of the recording jars or milk meters. It is normal for the sum of individual yields to be about 3% higher than bulk tank readings. This is owing to the cumulative effect of the tendency to read upwards when reading yields in the parlour.

Some farmers enter the cows' names on the sheets in the order in which they calved. This has a marginal advantage once a month when samples are taken for Official Milk Recording, but for the rest of the month it is unnecessarily time consuming to search for cow names, especially in large herds. Also, the dry cows may not, and seldom do, calve exactly on their expected date and can be entered in the wrong order. The preferred system of entering is alpha-numeric (or logical) and would read thus: Betty, Bugs, Joint, Mystery, R21, W79, 123, 128, 334.

Individual Animal Cards

Side 1: The calf record card, is used first viz: from birth to calving. The emphasis on this side is on getting the heifer to grow out well. The outlines are for Friesland and Ayrshire breeders to record markings (very useful if ear tags are lost). Heats and in seminations are recorded on side 2, and the card would normally be turned to this side once the first insemination has been done.

The Heifer Record Card (Side 1)

The complete fertility history is recorded here, and bull codes for use at first and subsequent in seminations are instantly accessible when needed. Side 2 flaps open to reveal sides 3 and, and on these two sides the complete production history of the cow can be recorded. Side 3 opens up to reveal sides 5 and 6. Side 5 is used to record the cow's physical and psychological attributes and is used in conjunction with the bull brochure in selecting suitable mates for the cow. Side 6 is used to record any health problems. Particular emphasis is on mastitis control.

These cards also are filed in alpha-numeric order so that they can be used conveniently. Further, their portability lends the system to field use. Coloured tags are available which can be clipped to the bottom of the card to signify, for example, cows to be pregnancy checked. The file or tray is carried in the shed and notes can be made directly onto the card. Different colours can, if desired, be used to indicate "dry", "pregnant" etc.

Lactation totals (up to 300 days and complete lactations) from the cow byre sheets are entered on these cards. If desired, these can be entered lightly in pencil and the "official" records superimposed when these are received.

These cards are not designed to be kept in shoe boxes, and proper file drawers or folders are well worth the cost involved.

Ring-binder Files and Plastic Filing Sleeves

In almost any herd the farmer collects pieces of paper relating to individual animals. The more common ones are breed society certificates and pedigrees, and Official Milk recording forms such as from LTD 1995 (the record of production and breeding performance received after each completed lactation). These forms, together with any other bits and pieces (photos etc.,), are most conveniently kept, again in alpha-numeric order, in transparent plastic filing sleeves in ring-binder files. Under this system the latest performance record (the old one having been discarded), or any other information can be referred to with the minimum of effort. A separate file/s should be used for culled or sold animals as these records are very seldom needed.

Cow and Heifer Calendars

Various "whole herd at a glance" methods are available and are essential ancillaries to the records listed above. In other words, these are not substitutes for the individual animal cards, but are supplementary to them, and are used to identify problem breeders, poor producers, cows due to show oestrus, and for other purposes. Each cow is represented by a two-or six-colour magnet which is placed on the board in a position and of a colour relevant to the latest event in her life. Full instructions on using these calendars are supplied by the manufacturers.

The two most useful makes presently available differ slightly, therefore their use will not be described in detail here. But the necessity for the calendar is its usefulness in highlighting actions and for planning decisions. Some examples of these are cows for pregnancy diagnosis (PD) and numbers of cows expected to calve each week for the next 6 months.

A heifer-rearing board can be constructed using "peg-board". This is marked similarly to the calf record card for mass-for-age graph, except that the calf outlines are omitted. Each heifer is represented by a wooden golf tee. Each month, when the heifers are weighed, the tees are moved to the correct mass and age position.

The background should be painted with green schoolboard paint and dosings and inoculations can be marked on the board at appropriate ages. Coloured washers (castrator rings are good) can be put onto the tees to record in seminations and pregnancy or even important dosings and inoculations, especially if a whole age group is not done simultaneously. There is also a commercially made heifer-rearing calendar using steel magnets. Unfortunately, it does not permit masses to be recorded.

Insemination/PD Notebook

Individual cow cards, unfortunately, do not lend themselves to certain important analyses and some duplication is unavoidable as will be clear from the following section on "Analysis of Herd Records". The insemination/PD notebook is simply a listing, in chronological order, of in seminations and their success or failure as determined at the regular veterinary inspection. A pocket-sized notebook which can be used for this purpose is available from an A.I. co-op.

The Daily Diary

A large format "page a day" diary should always be on the office table. This has many uses. In particular, it is used for recording events not recorded elsewhere. For example, Reserve the first few lines for appointments such as "Vet due at 9h00" and the rest for events.

"Dosed all calves in small shed with Lintex"; or "Started silage making today. 4 loads"; or "Top 60 cows to camp A1 (30 x 70 metres)" and so on. The diary is particularly important on large farms where more than one person may be involved in the management. The cow byre sheet can serve some of the daily diary functions, but not all.

Other Records

Files need to be maintained in the farm office for various bits of information which apply to the herd as a whole, such as leucocyte counts, milk recording summaries and T.B. tests.

It should be apparent that, if these records are going to be effectively kept and the maximum value extracted from them, a comfortable farm office is absolutely essential.

Analysis of Herd Records

No record keeping system is complete unless it leads to actions. Therefore the data collected and recorded in the sheets, cards, books and calendars must be processed (converted from data to information) before it is of real value.

Breeding and Selection

The individual cow cards lead directly to identification of poor producers, difficult breeders, mastitis problems etc. If the linear classification is used in conjunction with the A.I. cop-op's bull brochure, suitable mates and alternative mates can be chosen as described in KwaZulu-Natal Diary Leaflet 2.2 and recorded on the cow's card. In this way, when a cow is ready for insemination, last minute confusion over which bull to use on her will be avoided, as this will have been decided at leisure, perhaps in conjunction with an A.I. co-op or breed society field officer.

Action Lists

This terms refers to lists of names of animals requiring particular attention. The more important lists are discussed below.

Daily Action Lists

Cows likely to show heat: These can be identified directly from the up-to-date cow calendar. Cows within 18 to 24 days of their last heat or insemination should be actively sought out. It is quite likely that a cow in season will be missed when generally looking for heat signs in the herd as a whole. Small signs of heat such as a bull string or a roughened tail head can be picked up by having a good look at each cow on the 18 to 24 day list, especially in the morning, because night-time activity could be, and usually is missed. Such close examination of every animal in the herd is neither practicable nor necessary.

Cows due for drying-off: Too long a dry period is money wasted. Therefore higher producers must not be dried off too soon or too late. Once a cow has been confirmed in calf, her dry-off date can be entered on her card and at the start of each month dry off dates can be marked on the cow byre sheet. The cow calendar is not accurate enough to be used as the primary source of drying off information. Use the calving calendar of Table 1 to work out the drying-off dates, and enter these on the cow byre sheet so that cows are dried off on the correct day.

Cows and heifers due to calve: *Animals which are expected to calve must be carefully watched and any problems must receive attention.*

The cow calendar can also be used to generate this list. One should expect cows to calve up to a week early or late. Therefore, cows should appear on this list from seven days prior to the expected calving date, and remain on it until they calve.

Weekly Action Lists

Weekly activities will include dosings of younger animals, changes in feeding groups of growing animals and of cows, heifers now ready for insemination etc. Much of this can be read off the cow and heifer calendars. The alternative is regular perusal of the cards and the compiling of lists for the dairy.

Monthly Action Lists

An essential monthly activity (possibly fortnightly in large herds) is the routine veterinary examination. The following animals must be identified for this monthly visit:-

- Cows and heifers for pregnancy (PD) testing, i.e. cows which have gone 42 days from last insemination without recycling (list from the cow calendar). Put heifers onto the calendar when first inseminated. Use, say, the outer section of the circle so that they can be identified easily and not confused with the cows. For large herds of more than 150 cows, buy two calendars and put heifers and first-calvers on one and the rest of the herd on the other.

- Cows for speculum examination:-*All cows calved or aborted within the last month.*
- Cows for speculum re-examination:-*All cows treated (by uterine installation,* i.e. *douched) by vet. or farmer within the last month.*
- Cows which have gone 42 days from calving without having been reported on heat. These cows will retain their calving colour on their cow calendar magnets, amongst the rest whose magnets will have been turned to a different colour to reflect that they have shown heat.
- Cows with open periods exceeding 110 days and still not in calf. (Open period = period from calving to successful A.I.). They will stand out because their cow calendar magnets will be a different colour to the other cows whose magnet colour will reflect their status as in-calf. Reserve part of the calendar, say the centre area, for barren cows (cows on the cull list) so that they do not keep appearing in the action lists, or remove them from the active area of the board and put them in the top left hand corner.
- Cows with irregular cycles *(listed from the cow cards and A.I. book).*
- Cows confirmed in calf and which have cycled again.
- Any other problems.

Fertility Analysis

The pen and paper system described (or any other pen and paper system) requires a lot of hard work to produce herd summaries. These are critically important, as discussed in KwaZulu-Natal Diary Leaflet 2.3. Therefore, despite the hard work, a certain minimum number of analyses are too important to avoid. This is the one area where computers are of real value. All other aspects which have been discussed so far in this chapter do not require a computer for efficiency. In fact, for most of the information, a computer could be less efficient than pen and paper methods. For example, by the time the computer owner has loaded the right program, found the right cow, and found out which bull to use to inseminate her, the cow card user will be home drinking coffee, having completed his insemination. This example is only slightly exaggerated!

Inter calving period (ICP)

Mean ICP = exp {(In BP1 + ... + In BPn) / n} + 280 + VWP

where: BP = breeding period = ICP - VWP - 280

n = number of cows

VWP = voluntary waiting period

The best way to check the length of the VWP is to use the open period of the cow with the second shortest ICP. In other words, assume that the cow with the shortest ICP was bred too soon by mistake (maybe the neighbour's bull got to her) and use the days open of the next shortest as the indicator of the VWP. If it is shorter than expected then someone has been inseminating too soon after calving.

Services (SPC) and Total (TSPC) services per conception.

SPC = No. of in seminations to pregnant cows

No. of pregnant cows

TSPC = No. of in seminations performed

No. of calves born

TSPC data are not easy to collect. Some computer programs use a mathematical technique to estimate the TSPC. The method is not practicable without a computer. Therefore, the usual figure quoted is services per conception (SPC) based on pregnant cows only, a difference which could be significant. For example, the cows which fall pregnant could do so very easily but this could be only a small percentage of the herd, thus SPC on its own could mask a real problem in the rest of the herd.

Percentage Heats Spotted

"Unless we see our cows in season the best A.I. technique with the most fertile cows and bulls won't get our cows pregnant".

% Heats spotted = (SPC x 2100)/(ICP-VWP-270)

where: SPC = services per conception (pregnant cows only).

Q-sum Analysis

The Q-sum is a simple method of monitoring successes and failures and making comparisons between seasons, inseminators or even different bulls as well as identifying trends. A running Q-sum should be kept on the wall of the farm office and brought up to date at each regular veterinary visit. No other analysis gives an earlier warning of fertility problems.

The method is very simple. Assume that on a particular farm we were able to collect a list of insemination dates; who did the A.I., and the results of subsequent pregnancy diagnosis. These data could probably be read directly from the farmer's usual records, but they must be in chronological order so that trends can be detected. The data are then plotted as set out, if an event was successful it is plotted above the zero line and vice versa; each successive event is added to, or subtracted from, the one before.

The trick is to find the reasons for changes in trend, and in the rate of incline or decline, was it a change of season, a new batch of semen etc.? In large herds with several inseminators, keep Q-sums for each inseminator as well as for the herd as a whole. The big advantage of the Q-sum is that warnings are sounded within about two months of a change occurring. This will be a far better indicator that something is going wrong than with the usual calculations of average inter calving periods, days open, or in seminations per conception. These other parameters should still be calculated because they are important indicators.

If your reaction is: "What more records!"; then stop recording something less important. For example, how much use do you really make of all that pedigree information so patiently recorded back to the great grandsire on every cow's card?

Computerised Dairy Herd Records

It has become very fashionable to computerise dairy her record keeping. A common result is a mountain of printout which is only looked at when an adviser calls. Many enthusiastic computer users have never tried a good pen and paper system. The writer is an enthusiastic computer user and has one personal computer in his office and another at home. Nevertheless, it is his firm conviction that, for many farmers, the pen and paper system described is the better system and that, unless a good pen and paper system is thoroughly mastered, no farmer or adviser is in a position to evaluate a computerised record keeping system. Too many computer systems can pull out extended pedigrees which could be looked up more quickly in a good filing system, but do not even calculate inter calving periods for the herd or if they do, do it incorrectly.

The only area where a computer system is superior to a well designed pen and paper and cow calendar system is in the production of action lists in and the analyses described in KwaZulu-Natal Dairy leaflet 2.3, the farmer intending to buy a computerised record keeping program must ensure that features such as a mass-for-age analysis in the heifers, predicted future daily milk production and future calvings (very useful for cash flow budgeting) are included on the program. An on-farm computer system should be able :

- link up with a feeding program
- link to out of parlour feeders
- produce expected milk yields for the next milking
- and other practical outputs.

It is a simple truth that computers do not make life easier. In fact, they put more demands on management and, especially, do not take the place of competent advisers. For these reasons, many farmers are better served by not owning their own computers, but by making use of a pen and paper system for everyday management and by subscribing to a bureau service for analyses and feeding recommendations.

5

Milk Microbiology

Milk is a white liquid produced by the mammary glands of mammals. It provides the primary source of nutrition for young mammals before they are able to digest other types of food. The early lactation milk is known as colostrum, and carries the mother's antibodies to the baby. It can reduce the risk of many diseases in the baby. The exact components of raw milk vary by species, but it contains significant amounts of saturated fat, protein and calcium as well as vitamin C. Cow's milk has a pH ranging from 6.4 to 6.8, making it slightly acidic.

Types of Consumption

There are two distinct types of milk consumption: a natural source of nutrition for all infant mammals and a food product for humans of all ages that is derived from other animals.

Nutrition for Infant Mammals

In almost all mammals, milk is fed to infants through breastfeeding, either directly or by expressing the milk to be stored and consumed later. Some cultures, historically or currently, continue to use breast milk to feed their children until they are seven years old.

Human infants sometimes are fed fresh goat milk. There are known risks in this practice, including those of developing electrolyte imbalances, metabolic acidosis, megaloblastic anemia, and a host of allergic reactions.

Food Product for Humans

In many cultures of the world, especially the Western world, humans continue to consume milk beyond infancy, using the milk of other animals (especially cattle, goats and sheep) as a food product.

For millennia, cow's milk has been processed into dairy products such as cream, butter, yogurt, kefir, ice cream, and especially the more durable and easily transportable product, cheese. Modern industrial processes produce casein, whey protein, lactose, condensed milk, powdered milk, and many other food-additive and industrial products.

Humans are an exception in the natural world for consuming milk past infancy, despite the fact that many humans show some degree (some as little as 5%) of lactose intolerance, a characteristic that is more prevalent among individuals of African or Asian descent. The sugar lactose is found only in milk, forsythia flowers, and a few tropical shrubs.

The enzyme needed to digest lactose, lactase, reaches its highest levels in the small intestines after birth and then begins a slow decline unless milk is consumed regularly. On the other hand, those groups who do continue to tolerate milk often have exercised great creativity in using the milk of domesticated ungulates, not only of cattle, but also sheep, goats, yaks, water buffalo, horses, reindeers and camels. The largest producer and consumer of cattle and buffalo milk in the world is India.

Table 1: Top ten per capita cow's milk and cow's milk products consumers in 2006

Country	*Milk (liters)*	*Cheese (kg)*	*Butter (kg)*
Finland	183.9	19.1	5.3
Sweden	145.5	18.5	1.0
Ireland	129.8	10.5	2.9
Netherlands	122.9	20.4	3.3
Norway	116.7	16.0	4.3
Spain	119.1	9.6	1.0
Switzerland	112.5	22.2	5.6
United Kingdom	111.2	12.2	3.7
Australia	106.3	11.7	3.7
Canada	94.7	12.2	3.3

Terminology

The term *milk* is also used for white coloured, non-animal beverages resembling milk in colour and texture such as soy milk, rice milk, almond milk, and coconut milk. Even a substance regurgitated by pigeons to feed their young is called crop milk, though it bears little resemblance to mammalian milk.

Evolution

The mammary gland is thought to have been derived from apocrine skin glands. It has been suggested that the original function of lactation (milk production) was keeping eggs moist. Much of the argument is based on monotremes (egg-laying mammals): The original adaptive significance of milk secretions may have been nutrition or immunological protection. This secretion gradually became more copious and accrued nutritional complexity over evolutionary time.

History

Animal milk is known to have been used first as human food during the Secondary Products Revolution, around 5000 BC. It is presumed that when animals such as cattle were first domesticated, it was only for purposes of meat. Dairy products obtained from the animals proved to be a more efficient way of turning uncultivated grasslands into sustenance: the food value of an animal killed for meat can be matched by perhaps one year's worth of milk from the same animal, which will keep producing milk—in convenient daily portions—for years.

Milk by-products found inside Stone Age pottery from Turkey indicate processed milk was consumed in 6500 BC, some thousands of years before it is thought that adult humans had evolved the ability to digest raw milk.

DNA evidence extracted from Neolithic skeletons indicates that in 5500 BC, people in Northern Europe, as all other peoples of the time, were still lactose intolerant. Earthenware vessels found in England and dated to 4500 BC contain milk by-products, indicating milk was used in some form, although perhaps not drunk directly.

In 1863, French chemist and biologist Louis Pasteur invented pasteurization, a method of killing harmful bacteria in beverages and food products.

In 1884, Doctor Hervey Thatcher, an American inventor from New York, invented the first glass milk bottle, called 'Thatcher's Common Sense Milk Jar', which was sealed with a waxed paper disk. Later, in 1932, plastic-coated paper milk cartons were introduced commercially as a consequence of their invention by Victor W. Farris.

The town of Harvard, Illinois celebrates milk with a summer festival known as "Milk Days". Theirs is a different tradition meant to celebrate dairy farmers in the "Milk Capital of the World."

Sources

In addition to cattle, the following livestock provide milk used by humans for dairy products:

- Camel
- Donkey
- Goat
- Horse
- Reindeer
- Sheep
- Water buffalo
- Yak.

In Russia and Sweden, small moose dairies also exist.

According to the National Bison Association, American bison (also called American buffalo) are not milked commercially, however, various sources report cows resulting from cross-breeding bison and domestic cattle are good milk producers, and have been used both during the European settlement of North America and during the development of commercial Beefalo in the 1970s and 1980s.

Human milk is not produced or distributed industrially or commercially, however, milk banks exist that allow for the collection of donated human milk and its redistribution to infants who may benefit from human milk for various reasons (premature neonates, babies with allergies, metabolic diseases, etc.)

All other female mammals do produce milk, but it rarely or never is used to produce dairy products for human consumption.

Modern Production

In the Western world today, cow's milk is produced on an industrial scale and is by far the most commonly consumed form of milk. Commercial dairy farming using automated milking equipment produces the vast majority of milk in developed countries. Dairy cattle such as the Holstein have been bred selectively for increased milk production. About 90% of the dairy cows in the United States and 85% in Great Britain are Holsteins. Other dairy cows in the United States include Ayrshire, Brown Swiss, Guernsey, Jersey, and Milking Shorthorn (Dairy Shorthorn).

The largest producers of dairy products and milk today are India followed by the United States, Germany, and Pakistan.

Table 2: Top ten buffalo milk producers in 2007

Country	Production (tonnes)	Note
India	59,210,000	Unofficial/Semi-official/mirror data
Pakistan	20,372,000	official figure
People's Republic of China	2,900,000	FAO estimate
Egypt	2,300,000	
Nepal	958,603	official figure
Iran	241,500	FAO estimate
Myanmar	220,462	official figure
Italy	200,000	FAO estimate
Vietnam	32,000	
Turkey	30,375	official figure
World	86,574,539	Aggregate

Increasing affluence in developing countries, as well as increased promotion of milk and milk products, has led to a rise in milk consumption in developing countries in recent years. In turn, the opportunities presented by these growing markets have attracted investment by multinational dairy firms. Nevertheless, in many countries production remains on a small scale and presents significant opportunities for diversification of income sources by small farmers.

Local milk collection centers, where milk is collected and chilled prior to being transferred to urban dairies, are a good example of where farmers have been able to work on a cooperative basis, particularly in countries such as India.

The table below shows the numbers for water buffalo milk production. Cattle milk is produced in a much wider range.

Grading

United States

In the United States, there are two grades of milk, with Grade A primarily used for direct sales and consumption in stores, and Grade B used for indirect consumption, such as in cheese making or other processing.

The differences between the two grades are defined in the Wisconsin administrative code for Agriculture, Trade, and Consumer Protection, chapter 60. Grade B generally refers to milk that is cooled in milk cans, which are immersed in a bath of cold flowing water, that typically is drawn up from an underground water well rather than using mechanical refrigeration.

- Grade A farms are inspected every six months, while Grade B farms are inspected every two years {WI-ATCP 60.24.2}
- Both types of farms are required to have two cleaning vats in the milk house for washing and rinsing of equipment {WI-ATCP 60.07.2 (g)}. A farm also must have an additional separate sink and faucet provided for hand washing {WI-ATCP 60.07.2(h)}, unless the bulk tank was installed before Jan 1, 1979 or the farm uses milk cans.
- Grade A milk stored in a bulk tank is cooled to 45 degrees F within two hours of milking. Grade A milk in a tank may only rise to 50 F if milk from additional milking sessions is added to the tank (potentially requiring a plate cooler to reduce the temperature of a large volume influx quickly enough) and must be cooled back to 45 F within two hours. {WI-ATCP 60.2.4(b)}
- Grade B milk in milk cans is cooled to 50 degrees F within two hours of milking. Grade B farms cannot mix milk into cans from previous milking. {WI-ATCP 60.2.4(c)}
- The somatic cell count (SCC) of Grade A or B cow or sheep milk may not exceed 750,000 cells per mL, and the SCC of Grade A or B goat milk may not exceed 1,000,000 cells per mL. {WI-ATCP 60.15.4}
- The bacterial plate or loop count of Grade A milk may not exceed 100,000 per mL, while Grade B milk may not exceed 300,000 per mL. {WI-ATCP 60.15.2}
- A bacterial plate count test is required at least once a month. {WI-ATCP 60.18.3} If the bacterial count exceeds 100,000 per mL for Grade A or 300,000 per mL for grade B in 3 out of 5 tests, the license to sell milk is suspended. The license will be revoked immediately if the bacterial count ever exceeds 750,000 per mL. {WI-ATCP 60.18.6}

Price

It was reported in 2007 that with increased worldwide prosperity and the competition of bio-fuel production for feed stocks, both the demand for and the price of milk had substantially increased world wide. Particularly notable was the rapid increase of consumption of milk in China and the rise of the price of milk in the United States above the government subsidized price.

United States

In 2010 the Department of Agriculture predicted farmers would receive an average of $1.35 per U.S. gallon of cow's milk (35 cents per liter), which is down 30 cents per gallon from 2007 and below the break-even point for many cattle farmers.

Physical and Chemical Structure

Milk is an emulsion or colloid of butterfat globules within a water-based fluid.

Butterfat

Each fat globule is surrounded by a membrane consisting of phospholipids and proteins; these emulsifiers keep the individual globules from joining together into noticeable grains of butterfat and also protect the globules from the fat-digesting activity of enzymes found in the fluid portion of the milk. In unhomogenized cow's milk, the fat globules average about four micrometers across. The fat-soluble vitamins A, D, E, and K are found within the milk fat portion of the milk.

Other Proteins

The largest structures in the fluid portion of the milk are casein protein micelles: aggregates of several thousand protein molecules, bonded with the help of nanometer-scale particles of calcium phosphate. Each micelle is roughly spherical and about a tenth of a micrometer across. There are four different types of casein proteins, and collectively they make up around 80% of the protein in milk, by weight.

Most of the casein proteins are bound into the micelles. There are several competing theories regarding the precise structure of the micelles, but they share one important feature: the outermost layer consists of strands of one type of protein, k-casein, reaching out from the body of the micelle into the surrounding fluid. These kappa-casein molecules all have a negative electrical charge and therefore repel each other, keeping the micelles separated under normal conditions and in a stable colloidal suspension in the water-based surrounding fluid.

Milk contains dozens of other types of proteins beside the caseins. They are more water-soluble than the caseins and do not form larger structures. Because these proteins remain suspended in the whey left behind when the caseins coagulate into curds, they are collectively known as *whey proteins*. Whey proteins make up approximately 20% of the protein in milk, by weight. Lactoglobulin is the most common whey protein by a large margin.

Carbohydrates

The carbohydrate lactose gives milk its sweet taste and contributes approximately 40% of whole cow's milk's calories. Lactose is a disaccharide composite of two simple sugars, glucose and galactose. In nature, lactose is found only in milk and a small number of plants. Other components found in raw cow's milk are living white blood cells, mammary gland cells, various bacteria, and a large number of active enzymes.

Appearance

Both the fat globules and the smaller casein micelles, which are just large enough to deflect light, contribute to the opaque white colour of milk.

The fat globules contain some yellow-orange carotene, enough in some breeds (such as Guernsey and Jersey cattle) to impart a golden or "creamy" hue to a glass of milk. The riboflavin in the whey portion of milk has a greenish colour, which sometimes can be discerned in skimmed milk or whey products. Fat-free skimmed milk has only the casein micelles to scatter light, and they tend to scatter shorter-wavelength blue light more than they do red, giving skimmed milk a bluish tint.

Processing

In most Western countries, centralized dairy facilities process milk and products obtained from milk (dairy products), such as cream, butter, and cheese. In the U.S., these dairies usually are local companies, while in the Southern Hemisphere facilities may be run by very large nationwide or trans-national corporations (such as Fonterra).

Pasteurization

Pasteurization is used to kill harmful microorganisms by heating the milk for a short time and then cooling it for storage and transportation. Pasteurized milk still is perishable, however, and must be stored cold by both suppliers and consumers. Dairies print expiration dates on each container, after which stores will remove any unsold milk from their shelves. The process destroys the vitamin C content of the raw milk.

A newer process, ultrapasteurization or ultra-high temperature treatment (UHT), heats the milk to a higher temperature for a shorter amount of time. This extends its shelf life and allows the milk to be stored unrefrigerated because of the longer lasting sterilization effect, but it affects the taste adversely.

Microfiltration

Microfiltration is a process that partially replaces pasteurization and produces milk with fewer microorganisms and longer shelf life without a change in the taste of the milk. In this process, cream is separated from the whey and is pasteurized in the usual way, but the whey is forced through ceramic microfilters that trap 99.9% of microorganisms in the milk (as compared to 95% killing of microorganisms in conventional pasteurization). The whey then is recombined with the pasteurized cream to reconstitute the original milk composition.

Creaming and Homogenization

Upon standing for 12 to 24 hours, fresh milk has a tendency to separate into a high-fat cream layer on top of a larger, low-fat milk layer. The cream often is sold as a separate product with its own uses; today the separation of the cream from the milk usually is accomplished rapidly in centrifugal cream separators. The fat globules rise to the top of a container of milk because fat is less dense than water. The smaller the globules, the more other molecular-level forces prevent this from happening. In fact, the cream rises in cow's milk much more quickly than a simple model would predict: rather than isolated globules, the fat in the milk tends to form into clusters containing about a million globules, held together by a number of minor whey proteins. These clusters rise faster than individual globules can. The fat globules in milk from goats, sheep, and water buffalo do not form clusters so readily and are smaller to begin with; cream is very slow to separate from these milks.

Milk often is homogenized, a treatment which prevents a cream layer from separating out of the milk. The milk is pumped at high pressures through very narrow tubes, breaking up the fat globules through turbulence and cavitation. A greater number of smaller particles possess more total surface area than a smaller number of larger ones, and the original fat globule membranes cannot completely cover them. Casein micelles are attracted to the newly exposed fat surfaces; nearly one-third of the micelles in the milk end up participating in this new membrane structure. The casein weighs down the globules and interferes with the clustering that accelerated separation. The exposed fat globules are vulnerable to certain enzymes present in milk, which could break down the fats and produce rancid flavours. To prevent this, the enzymes are inactivated by pasteurizing the milk immediately before or during homogenization.

Homogenized milk tastes blander, but feels creamier in the mouth than unhomogenized; it is whiter and more resistant to developing off flavours. Creamline (or cream-top) milk is unhomogenized; it may or may not have been pasteurized. Milk which has undergone high-pressure homogenization, sometimes labelled as "ultra-homogenized," has a longer shelf life than milk which has undergone ordinary homogenization at lower pressures. Homogenized milk may be more digestible than unhomogenized milk.

Kurt A. Oster, M.D., who worked during the 1960s through the 1980s, suggested a link between homogenized milk and arterosclerosis, due to damage to plasmalogen resulting from the release of bovine xanthine oxidase (BXO) from the milk fat globular membrane (MFGM) during homogenization. Oster's hypothesis has been widely criticized, however, and has not been generally accepted by the scientific community. No link has been found between arterosclerosis and milk consumption.

Nutrition and Health

The composition of milk differs widely among species. Factors such as the type of protein; the proportion of protein, fat, and sugar; the levels of various vitamins and minerals; and the size of the butterfat globules, and the strength of the curd are among those than may vary. For example:

- Human milk contains, on average, 1.1% protein, 4.2% fat, 7.0% lactose (a sugar), and supplies 72 kcal of energy per 100 grams.
- Cow milk contains, on average, 3.4% protein, 3.6% fat, and 4.6% lactose, 0.7% minerals and supplies 66 kcal of energy per 100 grams. Donkey and horse milk have the lowest fat content, while the milk of seals and whales may contain more than 50%

fat. High fat content is not unique to aquatic mammals. Guinea pig milk has an average fat content of 46%.

Table 3: Milk composition analysis, per 100 grams

Constituents	Unit	Cow	Goat	Sheep	Waterbuffalo
Water	g	87.8	88.9	83.0	81.1
Protein	g	3.2	3.1	5.4	4.5
Fat	g	3.9	3.5	6.0	8.0
Carbohydrate	g	4.8	4.4	5.1	4.9
Energy	kcal	66	60	95	110
Energy	kJ	275	253	396	463
Sugars (lactose)	g	4.8	4.4	5.1	4.9
Cholesterol	mg	14	10	11	8
Calcium	mg	120	100	170	195
Saturated fatty acids	g	2.4	2.3	3.8	4.2
Monounsaturated fatty acids	g	1.1	0.8	1.5	1.7
Polyunsaturated fatty acids	g	0.1	0.1	0.3	0.2

Cow's Milk

These compositions vary by breed, animal, and point in the lactation period.

Table 4: Milk fat Percentages

Cow breed	Approximate percentage
Jersey	5.2
Zebu	4.7
Brown Swiss	4.0
Holstein-Friesian	3.6

The protein range for these four breeds is 3.3% to 3.9%, while the lactose range is 4.7% to 4.9%. Milk fat percentages may be manipulated by dairy farmers' stock diet formulation strategies. Mastitis infection can cause fat levels to decline.

Nutritional Value

Processed cow's milk was formulated to contain differing amounts of fat during the 1950s. One cup (250 ml) of 2%-fat cow's milk contains 285 mg of calcium, which represents 22% to 29% of the daily recommended intake (DRI) of calcium for an adult. Depending on the age, milk contains 8 grams of protein, and a number of other nutrients (either naturally or through fortification) including:

- Biotin
- Iodine
- Magnesium
- Pantothenic acid
- Potassium
- Riboflavin
- Selenium
- Thiamine
- Vitamin A
- Vitamin B_{12}
- Vitamins D
- Vitamin K.

The amount of calcium from milk that is absorbed by the human body is disputed.

Table 5: Cow's milk (whole) Nutritional value per 100 g (3.5 oz)

Energy	***252 kJ (60 kcal)***
Carbohydrates	5.26 g
Sugars	5.26 g
Lactose	5.26 g
Fat	3.25 g
saturated	1.865 g
monounsaturated	0.812 g
polyunsaturated	0.195 g
Protein	3.22 g
Tryptophan	0.075 g
Threonine	0.143 g
Isoleucine	0.165 g
Leucine	0.265 g
Lysine	0.140 g
Methionine	0.075 g
Cystine	0.017 g
Phenylalanine	0.147 g
Tyrosine	0.152 g
Valine	0.192 g

Arginine	0.075 g
Histidine	0.075 g
Alanine	0.103 g
Aspartic acid	0.237 g
Glutamic acid	0.648 g
Glycine	0.075 g
Proline	0.342 g
Serine	0.107 g
Water	88.32 g
Vitamin A equiv.	28 ìg (3%)
Thiamine (Vit. B_1)	0.044 mg (3%)
Riboflavin (Vit. B_2)	0.183 mg (12%)
Vitamin B_{12}	0.44 ìg (18%)
Vitamin D	40 IU (10%)
Calcium	113 mg (11%)
Magnesium	10 mg (3%)
Potassium	143 mg (3%)

100 mL corresponds to 103 g. Percentages are relative to US recommendations for adults. Source: USDA Nutrient database

Calcium from dairy products has a greater bioavailability than calcium from certain vegetables, such as spinach, that contain high levels of calcium-chelating agents, but a similar or lesser bioavailability than calcium from low-oxalate vegetables such as kale, broccoli, or other vegetables in the *Brassica* genus.

Medical Research

A 2006 study found that for women desiring to have a child, those who consume full fat dairy products may slightly increase their fertility, while those consuming low fat dairy products may slightly reduce their fertility.

Numerous studies have found that conjugated linoleic acid, found mainly in milk, meat and dairy products, provides several health benefits including prevention of atherosclerosis, different types of cancer, and hypertension and improved immune function.

There is recent evidence suggesting consumption of milk is effective at promoting muscle growth and improving post exercise muscle recovery.

In 2010, scientists at the Harvard School of Public Health identified a substance in dairy fat, trans-palmitoleic acid, that may substantially reduce the risk of type 2 diabetes. The researchers examined participants who have been followed for 20 years in an observational study to evaluate risk factors for cardiovascular diseases in older adults. During follow-up it was found that individuals with higher circulating levels of trans-palmitoleic acid had a much lower risk of developing diabetes, with about a 60% lower risk among participants in the highest quintile (fifth) of trans-palmitoleic acid levels.

Lactose Intolerance

Lactose, the disaccharide sugar component of all milk must be cleaved in the small intestine by the enzyme lactase in order for its constituents, galactose and glucose, to be absorbed. The production of this enzyme declines significantly after weaning in all mammals. Consequently, many humans become unable to digest lactose properly as they mature. There is a great deal of variance, with some individuals reacting badly to even small amounts of lactose, some able to consume moderate quantities, and some able to consume large quantities of milk and other dairy products without problems. When an individual consumes milk without producing sufficient lactase, they may suffer diarrhea, intestinal gas, cramps and bloating, as the undigested lactose travels through the gastrointestinal tract and serves as nourishment for intestinal microflora who excrete gas, a process known as anaerobic respiration.

It is estimated that 30 to 50 million Americans are lactose intolerant, including 75% of Native Americans and African Americans, and 90% of Asian Americans. Lactose intolerance is less common among those descended from northern Europeans.

Lactose intolerance is a natural process and there is no reliable way to prevent or reverse it.

Controversy

Some studies suggest that milk consumption may increase the risk of suffering from certain health problems. Cow milk allergy (CMA) is an immunologically mediated adverse reaction to one or more cow's milk proteins. Rarely is it severe enough to cause death.

Milk contains casein, a substance that breaks down in the human stomach to produce casomorphin, an opioid peptide. In the early 1990s it was hypothesized that casomorphin can cause or aggravate autism spectrum disorders, and casein-free diets are widely promoted. Studies supporting these claims have had significant flaws, and the data are inadequate to guide autism treatment recommendations.

A study demonstrated that men who drink a large amount of milk and consume dairy products were at a slightly increased risk of developing Parkinson's disease; the effect for women was smaller. The reason behind this is not fully understood, and it also remains unclear why there is less of a risk for women. Several sources suggest a correlation between high calcium intake (2000 mg per day, or twice the U.S. recommended daily allowance, equivalent to six or more glasses of milk per day) and prostate cancer. A large study specifically implicates dairy, i.e., low-fat milk and other dairy to which vitamin A palmitate has been added.

A review published by the World Cancer Research Fund and the American Institute for Cancer Research states that at least eleven human population studies have linked excessive dairy product consumption and prostate cancer.

Medical studies also have shown a possible link between milk consumption and the exacerbation of diseases such as Crohn's disease, Hirschsprung's disease–mimicking symptoms in babies with existing cow's milk allergies, and the aggravation of Behçet's disease.

Bovine growth Hormone Supplementation

Since November 1993, with FDA approval, Monsanto has been selling recombinant bovine somatotropin (rbST), also called rBGH, to dairy farmers. Cows produce bovine growth hormone naturally, but some producers administer an additional recombinant version of BGH which is produced through a genetically engineered E. coli because it increases milk production. Bovine growth hormone also stimulates liver production of insulin-like growth factor 1 (IGF1).

Monsanto has stated that both of these compounds are harmless given the levels found in milk and the effects of pasteurization, however, Monsanto's own tests, conducted in 1987, demonstrated that statistically significant growth stimulating effects were induced in organs of adult rats by feeding IGF-1 at low dose levels for only two weeks. "Drinking rBGH milk would thus be expected to significantly increase IGF-1 blood levels and consequently to increase risks of developing breast cancer and promoting its invasiveness."

On June 9, 2006, the largest milk processor in the world and the two largest supermarkets in the United States—Dean Foods, Wal-

Mart, and Kroger—announced that they are "on a nationwide search for rBGH-free milk." Milk from cows given rBST may be sold in the United States, and the FDA stated that no significant difference has been shown between milk derived from rBST-treated and that from non-rBST-treated cows. Milk that advertises that it comes from cows not treated with rBST, is required to state this finding on its label.

Cows receiving rBGH supplements may more frequently contract an udder infection known as mastitis. Problems with mastitis have led to Canada, Australia, New Zealand, and Japan banning milk from rBST treated cows. Mastitis, among other diseases, may be responsible for the fact that levels of white blood cells in milk vary naturally.

In the European Union, rBGH is banned.

Ethical Concerns

Vegans and some other vegetarians do not consume milk for a variety of reasons. They may object to features of dairy farming including the necessity of killing almost all the male offspring of dairy cows (either by disposal soon after birth, for veal production, or for beef), the routine separation of mother and calf soon after birth, other perceived inhumane treatment of dairy cattle, and culling of cows after their productive lives.

Flavoured Milk in U.S. Schools

According to an article in *The New York Times*, milk must be offered at every meal if a United States school district wishes to get reimbursement from the federal government. A quarter of the largest school districts in the U.S. offer rice or soy milk and almost 17% of all U.S. school districts offer lactose-free milk. Seventy-one percent of the milk served in the U.S. is flavoured, causing some school districts to propose a ban because flavoured milk has added sugars. The Boulder, Colorado school district banned flavoured milk in 2009 and instead installed a dispenser that keeps the milk colder.

Varieties and Brands

Milk products are sold in a number of varieties based on types/ degrees of;

- additives (e.g., vitamins),
- age (e.g., cheddar),
- coagulation (e.g., cottage cheese),
- farming method (e.g., organic, grass-fed).

- fat content (e.g., half and half),
- fermentation (e.g., buttermilk),
- flavoring (e.g., chocolate),
- homogenization (e.g., cream top),
- mammal (e.g., cow, goat, sheep),
- packaging (e.g., bottle),
- pasteurization (e.g., raw milk),
- water content (e.g., dry milk).

Milk preserved by the UHT process does not need to be refrigerated before opening and has a longer shelf life than milk in ordinary packaging. It is typically sold unrefrigerated in Europe, Latin America, and Australia.

Additives and Flavoring

In areas where the cattle (and often the people) live indoors, commercially sold milk commonly has vitamin D added to it to make up for lack of exposure to UVB radiation.

Reduced fat milks often have added vitamin A palmitate to compensate for the loss of the vitamin during fat removal; in the United States this results in reduced fat milks having a higher vitamin A content than whole milk.

To aid digestion in those with lactose intolerance, milk with added bacterial cultures such as *Lactobacillus acidophilus* ("acidophilus milk") and bifidobacteria ("a/B milk") is available in some areas. Another milk with *Lactococcus lactis* bacteria cultures ("cultured buttermilk") often is used in cooking to replace the traditional use of naturally soured milk, which has become rare due to the ubiquity of pasteurization, which also kills the naturally occurring lactococcus bacteria.

Milk often has flavoring added to it for better taste or as a means of improving sales. Chocolate milk has been sold for many years and has been followed more recently by strawberry milk and others. Some nutritionists have criticized flavoured milk for adding sugar, usually in the form of high-fructose corn syrup, to the diets of children who are already commonly obese in the U.S.

Distribution

Due to the short shelf-life of normal milk, it used to be delivered to households daily in many countries. However, improved refrigeration at home, changing food shopping patterns because of supermarkets, and the higher cost of home delivery, means that daily deliveries by a milkman is no longer available in most countries.

Australia and New Zealand

In Australia and New Zealand, prior to "metrification", milk was generally distributed in 1 pint (568ml) glass bottles. In Australia and in Ireland there was a government funded "free milk for school children" program, and milk was distributed at morning recess in 1/3 pint bottles. With the conversion to metric measures, the milk industry were concerned that the replacement of the pint bottles with 500ml bottles would result in a 13.6% drop in milk consumption.

Hence, all pint bottles were recalled and replaced by 600ml bottles. With time, due to the steadily increasing cost of collecting, transporting, storing and cleaning glass bottles, they were replaced by cardboard cartons.

A number of designs were used, including a tetrahedron which could be close-packed without waste space, and could not be knocked over accidentally. (slogan: No more crying over spilt milk.) However, the industry eventually settled on a design similar to that used in the United States. Milk is now available in a variety of sizes in cardboard cartons (250 mL, 375 mL, 600 mL, 1 liter and 1.5 liters) and plastic bottles (1, 2 and 3 liters).

A significant addition to the marketplace has been "long life" milk (UHT), generally available in 1 and 2 liter rectangular cardboard cartons. In urban and suburban areas where there is sufficient demand, home delivery is still available, though in suburban areas this is often 3 times per week rather than daily. Another significant and popular addition to the marketplace has been flavoured milks – for example, as mentioned above, Farmers Union Iced Coffee outsells Coca-Cola in South Australia.

India

In rural India, milk is delivered daily by a local milkman carrying bulk quantities in a metal container, usually on a bicycle; and in other parts of metropolitan India, milk is usually bought or delivered in a plastic bags or cartons via shops or supermarkets.

Pakistan

In Pakistan, milk is supplied in jugs. Milk has been a staple food, especially among the pastoral tribes in this country.

United Kingdom

Since the late 1990s, milk buying patterns have changed drastically in the UK. The classic milkman, who travels his local milk round (route) using a milk float (often battery powered) during the early hours and delivers milk in 1 pint glass bottles with aluminium foil tops directly to households, has almost disappeared. The main reasons for the decline of UK home deliveries by milkmen are household refrigerators (which lessen the need for daily milk deliveries) and private car usage (which has increased supermarket shopping).

In 1996, more than 2.5 billion liters of milk were still being delivered by milkmen, but by 2006 only 637 million liters (13% of milk consumed) was delivered by some 9,500 milkmen. By 2010, the estimated number of milkmen had dropped to 6,000. Assuming that delivery per milkman is the same as it was in 2006, this means milkmen deliveries now only account for 6–7% of all milk consumed by UK households (6.7 billion liters in 2008/2009). Almost 95% of all milk in the UK is thus sold in shops today, most of it in plastic bottles of various sizes, but some also in milk cartons. Milk is hardly ever sold in glass bottles in UK shops.

United States

In the United States, glass milk bottles have been replaced mostly with milk cartons and plastic jugs. Gallons of milk are almost always sold in jugs, while half-gallons and quarts may be found in both paper cartons and plastic jugs, and smaller sizes are almost always in cartons. Recently, milk has been sold in smaller resealable bottles made to fit in automobile cup holders.

The .5 US pints (240 ml; 8.3 imp fl oz) milk carton is the traditional unit as a component of school lunches, though some companies have replaced that unit size with a plastic bottle, which is also available at retail in 6 and 12-pack size. Ultrapasturized milk which does not need refrigeration or in containers which can hold the cold for as long as eight hours at room temperature are available.

Packaging

Glass milk bottles are now rare. Most people purchase milk in bags, plastic bottles, or plastic-coated paper cartons. Ultraviolet (UV) light from fluorescent lighting can alter the flavour of milk, so many companies that once distributed milk in transparent or highly translucent containers are now using thicker materials that block the UV light. Milk comes in a variety of containers with local variants:

- *Australia and New Zealand:* Distributed in a variety of sizes, most commonly in aseptic cartons for up to 1.5 liters, and plastic screw-top bottles beyond that with the following volumes; 1.1 L, 2 L, and 3 L. 1 liter milk bags are starting to appear in supermarkets, but have not yet proved popular. Most UHT-milk is packed in 1 or 2 liter paper containers with a sealed plastic spout.
- *Brazil:* Used to be sold in cooled 1 liter bags, just like in South Africa. Today the most common form is 1 liter aseptic cartons containing UHT skimmed, semi-skimmed or whole milk, although the plastic bags are still in use for pasteurized milk. Higher grades of pasteurized milk can be found in cartons or plastic bottles. Sizes other than 1 liter are rare.
- *Canada:* 1.33 liter plastic bags (sold as 4 liters in 3 bags) are widely available in some areas (especially the Maritimes, Ontario and Quebec), although the 4 liter plastic jug has supplanted them in western Canada. Other common packaging sizes are 2 liter, 1 liter, 500 mL, and 250 mL cartons, as well as 4 liter, 1 liter, 250 mL aseptic cartons and 500 mL plastic jugs.
- *Chile:* Distributed most commonly in aseptic cartons for up to 1 liter, but smaller, snack-sized cartons are also popular. The most common flavours, besides the natural presentation, are chocolate, strawberry and vanilla.
- *China:* Sweetened milk is a drink popular with students of all ages and is often sold in small plastic bags complete with straw. Adults not wishing to drink at a banquet often drink milk served from cartons or milk tea.
- Colombia sells milk in 1 liter plastic bags.
- Croatia, Bosnia and Herzegovina, Serbia, Montenegro.

  UHT milk is sold in 500 mL and 1 L (sometimes also 200 ml) aseptic cartons. Non-UHT pasteurized milk is most commonly sold in 1 L and 1.5 L PET bottles, though in Serbia one can still find milk in plastic bags.
- *Parts of Europe:* Sizes of 500 mL, 1 liter (the most common), 1.5 liters, 2 liters and 3 liters are commonplace.
- *Finland:* Commonly sold in 1 L or 1.5 L cartons, in some places also in 2 dl and 5 dl cartons.

- *Hong Kong:* Milk is sold in glass bottles (220 mL), cartons (236 mL and 1 L), plastic jugs (2 liters) and aseptic cartons (250 mL).
- *India:* Commonly sold in 500 mL plastic bags and in bottles in some parts like in west. It is still customary to serve the milk boiled, despite pasteurization. Milk is often buffalo milk. Flavoured milk is sold in most convenience stores in waxed cardboard containers. Convenience stores also sell many varieties of milk (such as flavoured and ultra-pasteurized) in different sizes, usually in aseptic cartons.
- *Indonesia:* Usually sold in 1 liter cartons, but smaller, snack-sized cartons are available.
- *Israel:* Non-UHT milk is most commonly sold in 1 liter waxed cardboard boxes and 1 liter plastic bags. It may also be found in 0.5 L and 2 L waxed cardboard boxes, 2 L plastic jugs and 1 L plastic bottles. UHT milk is available in 1 liter (and less commonly also in 0.25 L) carton "bricks".
- *Japan:* Commonly sold in 1 liter waxed paperboard cartons. In most city centers there is also home delivery of milk in glass jugs. As seen in China, sweetened and flavoured milk drinks are commonly seen in vending machines.
- *South Korea:* Sold in cartons (180 mL, 200 mL, 500 mL 900 mL, 1 L, 1.8 L, 2.3 L), plastic jugs (1 L and 1.8 L), aseptic cartons (180 mL and 200 mL) and plastic bags (1 L).
- *Pakistan:* Milk is supplied in 500 mL Plastic bags and carried in Jugs from rural to cities and sell
- *Poland:* UHT milk is mostly sold in aseptic cartons (500 mL, 1 L, 2 L), and non-UHT in 1 L plastic bags or plastic bottles. Milk, UHT is commonly boiled, despite being pasteurized.
- *South Africa:* Commonly sold in 1 liter bags. The bag is then placed in a plastic jug and the corner cut off before the milk is poured.
- *Sweden:* Commonly sold in 0.3 L, 1 L or 1.5 L cartons and sometimes as plastic or glass milk bottles.
- *Turkey:* Commonly sold in 500 mL or 1L cartons or special plastic bottles. UHT milk is more popular. Milkmen also serve in smaller towns and villages.
- *United Kingdom:* Most stores stock imperial sizes: 1 pint (568 mL, 2 pints (1.136 L), 4 pints (2.273 L), 6 pints (3.408 L)

or a combination including both metric and imperial sizes. Glass milk bottles delivered to the doorstep by the milkman are typically pint-sized and are returned empty by the householder for repeated reuse. Milk is sold at supermarkets in either aseptic cartons or HDPE bottles. Supermarkets have also now begun to introduce milk in bags, to be poured from a proprietary jug and nozzle.

- *United States:* Commonly sold in gallon (3.78 L), half-gallon (1.89 L) and quart (0.94 L) containers of natural-colored HDPE resin, or, for sizes less than one gallon, cartons of waxed paperboard. Bottles made of opaque PET are also becoming commonplace for smaller, particularly metric, sizes such as one liter. The U.S. single-serving size is usually the half-pint (about 240 mL). Less frequently, dairies deliver milk directly to consumers, from coolers filled with glass bottles which are typically half-gallon sized and returned for reuse. Some convenience store chains in the United States (such as Kwik Trip in the Midwest) sell milk in half-gallon bags, while another rectangular cube gallon container design used for easy stacking in shipping and displaying is used by warehouse clubs such as Costco and Sam's Club, along with some Wal-Mart stores.
- *Uruguay:* Commonly sold in 1 liter bags. The bag is then placed in a plastic jug and the corner cut off before the milk is poured.

Practically everywhere, condensed milk and evaporated milk is distributed in metal cans, 250 and 125 mL paper containers and 100 and 200 mL squeeze tubes, and powdered milk (skim and whole) is distributed in boxes or bags.

Spoilage and Fermented Milk Products

When raw milk is left standing for a while, it turns "sour". This is the result of fermentation, where lactic acid bacteria ferment the lactose in the milk into lactic acid. Prolonged fermentation may render the milk unpleasant to consume. This fermentation process is exploited by the introduction of bacterial cultures (e.g. *Lactobacilli sp., Streptococcus sp., Leuconostoc sp.*, etc) to produce a variety of fermented milk products. The reduced pH from lactic acid accumulation denatures proteins and causes the milk to undergo a variety of different transformations in appearance and texture, ranging from an aggregate to smooth consistency. Some of these products include sour cream, yoghurt, cheese, buttermilk, viili, kefir, and kumis.

Pasteurization of cow's milk initially destroys any potential pathogens and increases the shelf-life, but eventually results in spoilage that makes it unsuitable for consumption. This causes it to assume an unpleasant odour, and the milk is deemed non-consumable due to unpleasant taste and an increased risk of food poisoning. In raw milk, the presence of lactic acid-producing bacteria, under suitable conditions, ferments the lactose present to lactic acid. The increasing acidity in turn prevents the growth of other organisms, or slows their growth significantly. During pasteurization however, these lactic acid bacteria are mostly destroyed.

In order to prevent spoilage, milk can be kept refrigerated and stored between 1 and 4 degrees Celsius in bulk tanks. Most milk is pasteurized by heating briefly and then refrigerated to allow transport from factory farms to local markets. The spoilage of milk can be forestalled by using ultra-high temperature (UHT) treatment; milk so treated can be stored unrefrigerated for several months until opened but has an unpleasant "cooked" taste. Condensed milk, made by removing most of the water, can be stored in cans for many years, unrefrigerated, as can evaporated milk. The most durable form of milk is powdered milk, which is produced from milk by removing almost all water. The moisture content is usually less than 5% in both drum- and spray-dried powdered milk.

Language and Culture

The importance of milk in human culture is attested to by the numerous expressions embedded in our languages, for example "the milk of human kindness". In ancient Greek mythology, the goddess Hera spilled her breast milk after refusing to feed Heracles, resulting in the Milky Way. In African and Asian developing nations, butter is traditionally made from fermented milk rather than cream. It can take several hours of churning to produce workable butter grains from fermented milk.

Holy books have also mentioned milk; the Bible contains references to the 'Land of Milk and Honey'. In the Qur'an, there is a request to wonder on milk as follows: 'And surely in the livestock there is a lesson for you, We give you to drink of that which is in their bellies from the midst of digested food and blood, pure milk palatable for the drinkers.' (16-The Honeybee, 66). The Ramadhan fast is traditionally broken with a glass of milk and dates.

The verb, "to milk" something is often used in the vernacular of many English-speaking countries as a synonym for extortion or, in less loaded terms, taking advantage of a situation where one has another person at a disadvantage, as in 'milking the situation'.

The word milk has had many slang meanings over time. In the early 17th century the word was used to mean semen, or vaginal secretions, or to masturbate oneself or someone else. In the 19th century, milk was used to describe a cheap alcoholic drink made from methylated spirits mixed with water. The word was also used to mean defraud, to be idle, to intercept telegrams addressed to someone else, and a weakling or 'milksop'. In the mid 1930s, the word was used in Australia meaning to siphon gas from a car.

Milk is sometimes referred to as moo juice in American English, while Cockney rhyming slang calls it Acker Bilk, Tom Silk, Lady in silk and Kilroy Silk.

The name of the Russian Molokan religion in Russian is derived from Russian "Ìîëîêî " meaning "Milk" as they would drink milk on the Russian Orthodox days of fast.

Other Uses

Besides serving as a beverage or source of food, milk has been described as used by farmers and gardeners as an organic fungicide and foliage fertilizer. Diluted milk solutions have been demonstrated to provide an effective method of preventing powdery mildew on grape vines, while showing it is unlikely to harm the plant.

Microorganisms in Milk

The types of micro organisms found in milk vary considerably, and are dependent upon the specific conditions associated with a batch of milk.

Bacteria, yeasts, moulds, and bacteriophages are commonly encountered. Other viruses and Protozoa are seldom observed in milk products, except as occasional contaminants.

Bacteria

Bacteria are the most common, and probably the most numerous of microorganisms with which the dairy processing industry is concerned.

They belong to four main groups: (1) cocci, usually gram positive, (2) gram positive non-sporeforming rods, (3) gram positive sporeforming rods, and (4) gram negative non-spore forming rods.

Yeasts

The yeasts most frequently encountered in milk and milk products act upon the lactose to produce acid and carbon dioxide.

Yeasts are more commonly found in raw, cream during hot weather, but are potential contaminants throughout the year.

Moulds

Moulds often grow in large concentrations and are visible as a fuzzy or fluffy growth. They are sometimes observed on the surface, of butter, old cream, khoa, or cheese.

They are black, grey, green, blue or white. They discolour milk products and often produce undesir-able, at times repulsive, odours and flavours. Moulds are essential in production of the certain kinds of cheese.

Bacteriophages

Bacteriophages are particularly obnoxious in starter cultures used for making cultured milk, butter and cheese.

Phages will, kill the bacterial culture and the whole process of fermentation will be lost.

Microorganisms found in milk can also be described on the basis of the following characteristics;

1. Biochemical activities.
2. Temperature response.
3. Ability to cause infection and disease.

Biochemical Activities

If allowed to stand under condition that permit bacterial growth, raw milk of a good sanitary quality will rapidly undergo a series of chemical changes. The principal change is lactose fermentation to lactic acid. This change is brought about by acid uric lactic organisms, especially Strepotococcus lactis and certain lactobacilli. These include two distinct biochemical types, homo-and heterofermentative. In homofermentation lactic acid is the major product of lactose fermentation.

Heterofermentative organisms, however, produce lactic, acetic, propionic, and some other acids, and some alcohols and gases such as CO_2 and H_2 Organisms continue to form lactic acid until the concentration of acid is itself too great for the organisms to remain live.

Microbacteria, micrococci, coliforn, etc. also ferment lactose to lactic acid and other products. Many Clostriifiul1J species and, some yeasts such as Torula lactic, and Torula cremoris ferment lactose with acid and gas production.

As the acidity continues to increase and reaches a pH of 4.7, it eventually causes a precipitation of casein. Organisms capable of metabolizing lactic and other acids develop especially acid uric, yeasts and moulds.

The acidity of milk is diminished and the alkaline products of protein decomposition such as amines, ammonia and the like are produced.

This is accomplished by many species of the genera Bacillus, Clostridium, Pseudomonas, Proteus and numerous other forms.

The action of microorganisms does not involve fat as readily as it does lactose and protein. Lipolysis results from the action of lipase produced by bacteria such as Pseudomonas, Achromobacter and by some yeasts and moulds.

Fat is hydrolysed to glycerol and fatty acids. Some of the fatty acids, for example, butyric and caproic acid give milk products, distinctive and usually rancid, odours and flavours.

Several microorganisms also bring about certain objection able changes in the milk which may not be deleterious to health. Rapines in milk is sometimes encountered. The milk become ropy or slimy and may be pulled out into long threads.

It is produced by several organisms but the most important species is Alcaligenes viscolactis. A rapid fermentation of lactose in milk is sometimes observed and is known as stormy fermentation. This is brought about by Clostridium perfringens.

The curd become torn to shreds by the vigorous fermentation and gas production. Several organisms have been isolated from milk which impart brilliant colours. Pseudomonas syncyanea imparts blue colour, pseudomonas synxantha yellow colour and Serratia marcescens red colour to the milk.

Temperature Response

Microorganisms found in milk can also be described according their optimum temperature for growth and heat resistance. This is a very practical consideration since milk is preserved by employing low temperatures to prevent changes due to microbial activi1y' and by high temperatures to reduce microbial population and destroy pathogens. All the four types of microorganisms i.e. psychrophilic, mesophilic, thermophilic and thermoduric are found in milk.

Psychrophiles grow at temperatures just above freezing and at refrigeration tempratures. They produce a wide variety of spoilage defects. The defects may result in the production of many "off" flavours and odours. The most commonly encountered psychrophilic bacteria are members of the genera pseudomonas, Achromobacter, Vibrio, Flavobacterium and Alcaligenes, They arc killed in the pasteurization process, but are sometimes found in pasteurized milk. The contamination takes place after pasteurization from equipment, cans, bottles, and water.

The most important mesophilic bacteria are streptococci, lactobacilli and coliforms, which produce acid and gas and off flavours. They are killed in the pasteurization process

Thermophilic bacteria grow well at the temperature used in pasteurization, specially when the low temperature holding method is followed. Most thermophilic forms are found in two genera, Bacillus and Clostridium.

Thermoduric organisms are regarded as those which survive pasteurization but do not grow at pasteurization temperatures. The most common thermoduric bacteria are found in the genera Microbacterium, Corynebacterium, Micrococcus, Streptococcus and Bacillus.

Excessive numbers of thermoduric bacteria in milk make it difficult to meet the grading standard.

Ability to Cause Infection and Disease

Pathogenic organisms of both bovine and human origin have-been isolated from milk. Milk, therefore, can serve as a carrier of diseases. Many serious epidemics were caused by the consumption of such products before this fact was clearly recognized.

However, this became less common as milk sanitation has improved and pasteurization is being more widely practised. The disease organisms present in milk may be derived from (1) diseased animals or (2) persons collecting and handling milk: Thus the danger is due to the inoculum and not to the growth of organisms in the milk.

The health of animal is an important factor. Several diseases of cattle including staphylococcal and streptococcal infections, tuberculosis, brucellosis, salmonellosis, Q fever and Foot and mouth disease may be transmitted to man. The organisms causing these diseases may get into the milk either directly from the udder, or indirectly from infected body discharges, which may drop, splash, or be blown into the milk.

Some of the important diseases of human origin that have been transmitted by milk are (1) typhoid fever (2) diphtheria, (3) scarlet fever, (4) dysentery (5) septic sore throat and (6) poliomyelitis.

It is also possible for humans to infect animals. For example, mastitis may be caused by a variety of organisms, including Staphylococcus aureus. The infecting organism, in same cases, has been traced to humans.

Microbiology of Raw Milk

Raw milk is milk that has not been pasteurized or homogenized.

Humans consumed raw milk exclusively prior to the industrial revolution and the discovery of the pasteurization process in 1864. During the industrial revolution large populations congregated into urban areas detached from the agricultural lifestyle. Up until that point, individuals and families owned their own goats, cows, and other livestock and milked them on a daily basis.

Pasteurization was first used in the United States in the 1890s after the discovery of germ theory to control the hazards of highly contagious bacterial diseases including bovine tuberculosis and brucellosis that was thought to be easily transmitted to humans through the drinking of raw milk. Initially after the scientific discovery of bacteria, no product testing was available to determine if a farmer's milk was safe or infected, so all milk was treated as potentially contagious. After the first test was developed, some farmers actively worked to prevent their infected animals from being killed and removed from food production, or would falsify the test results so that their animals would appear to be free of infection.

When it was first used, pasteurization was thought to make raw milk from any source safer to consume. More recently, farm sanitation has greatly improved and effective testing has been developed for bovine tuberculosis and other diseases, making other approaches to ensuring safety of milk more feasible; however pasteurization continues to be widely used to prevent infected milk from entering the food supply. The recognition of many potentially deadly pathogens, such as E. Coli O157:H7, Listeria, and Salmonella, and their presence in milk products has led to the continuation of pasteurization. The Department of Health and Human Services, Centre for Disease Control and Prevention, and other health agencies of the United States strongly recommend that the public does not consume raw milk or raw milk products. Young children, the elderly, people with weakened immune systems, and pregnant women are particularly susceptible to infections originating in raw milk.

Recent advances in the analysis of milk-borne diseases have enabled scientists to track the DNA of the infectious bacteria to the cows on the farms that supplied the raw milk . This technique eliminates much of the debate of the merits of safe milk practices.

Raw vs. Pasteurized Debate

The raw vs. pasteurized debate pits the alleged health benefits of consuming raw milk against the disease threat of unpasteurized milk. Although agencies such as the Centers for Disease Control (CDC), the Food and Drug Administration (FDA), and some regulatory agencies around the world say that pathogens from raw milk make it unsafe to consume, some organizations say that raw milk can be produced hygienically, and that it has health benefits that are destroyed in the pasteurization process. This latter statement is not supported by all research.

Milk Spoilage

Spoilage occurs when microorganisms degrade the carbohydrates, proteins, fats of milk and produce noxious, end products.

It may be seen that Lactobacillus or Streptococcus species ferment the lactose to lactic acid and acetic acids turning the milk sour. They may produce enough acid to curdle the protein and form sour curd.

Attack of milk protein by Micrococcus, Bacillus or Proteus results into sweet curdling.

There is little acid formation. If milk becomes contaminated with Gram-negative rods-of coliform group of bacteria, such as E. coli or Enterobacter aerogenes, or Clostridium sp., there is .acid and gas formation from the lactose.

This stormy fermentation causes the explosion of curds.

Ropiness, like bread develops from Alcaligenes, Klebsiella and Enterobacter.

Serratia marcescens causes the development of a red pigment.

Milk-borne Diseases

The important diseases are, tuberculosis, brucellosis and Q fever. Tuberculosis bacterium. Mycobacterium bovis is consumed in the milk and passes from human intestine to the blood, from which it spreads to most organs.

Brucellosis, a blood-disease is caused by Gram-negative rod, Brucella abortus. When transmitted to man through cow milk, the bacterium infect the blood-rich organs. Q fever, caused by rickettsia, Cxiella burnetii is also a milk borne disease.

Other important disorders associated with milk are primary atypical pneumonia, toxoplasmosis, anthrax, streptococcal infections etc.

Milk Microorganisms Ability to Cause Infection and Diseases

Pathogenic organisms of both bovine and human origin have-been isolated from milk. Milk, therefore, can serve as a carrier of diseases.

Many serious epidemics were caused by the consumption of such products before this fact was clearly recognized.

However, this became less common as milk sanitation has improved and pasteurization is being more widely practised.

The disease organisms present in milk may be derived from;

(1) diseased animals or

(2) persons collecting and handling milk:

Thus the danger is due to the inoculum and not to the growth of organisms in the milk.

The health of animal is an important factor. Several diseases of cattle including staphylococcal and streptococcal infections, tuberculosis, brucellosis, salmonellosis,

Q fever and Foot and mouth disease may be transmitted to man.

The organisms causing these diseases may get into the milk either directly from the udder, or indirectly from infected body discharges, which may drop, splash, or be blown into the milk.

Some of the important diseases of human origin that have been transmitted by milk are;

(1) typhoid fever

(2) diphtheria,

(3) scarlet fever,

(4) dysentery

(5) septic sore throat and

(6) poliomyelitis. It is also possible for humans to infect animals.

For example, mastitis may be caused by a variety of organisms, including Staphylococcus aureus.

The infecting organism, in same cases, has been traced to humans.

Milk Allergy

Milk allergy is a food allergy, an adverse immune reaction to one or more of the proteins in cow's milk and/or the milk of other animals—proteins that are normally harmless to the non-allergic individual. Milk allergy can cause anaphylaxis, a potentially life-threatening condition. Milk allergy is a separate and distinct condition from lactose intolerance. (Lactose intolerance is considered the normal state for most adults on a worldwide scale and is not typically considered to be a disease condition.)

Symptoms

The principal symptoms are gastrointestinal, dermatological and respiratory. These can translate to: skin rash, hives, vomiting, and gastric distress such as diarrhea, constipation, stomach pain or flatulence. The clinical spectrum extends to diverse disorders: anaphylactic reactions, atopic dermatitis, wheeze, infantile colic, gastroesophageal reflux (GER), oesophagitis, allergic colitis headache/migraine, oral irritation, and constipation.

The symptoms may occur within a few minutes after exposure in *immediate reactions*, or after hours (and in some cases after several days) in *delayed reactions*.

Difference between Milk Allergy and Lactose Intolerance

Milk allergy is a food allergy, an adverse immune reaction to a food protein that is normally harmless to the non-allergic individual. Lactose intolerance is a non-allergic food sensitivity, and comes from a lack of production of the enzyme lactase, required to digest the predominant sugar in milk. Adverse effects of lactose intolerance generally occur after much higher levels of milk consumption than do adverse effects of milk allergy.

Difference from Milk Protein Intolerance

Milk protein intolerance (MPI) is delayed reaction to a food protein that is normally harmless to the non-allergic, non-intolerant individual. Milk protein intolerance produces a non-IgE antibody and is not detected by allergy blood tests. Milk protein intolerance produces a range of symptoms very similar to milk allergy symptoms, but can also include blood and/or mucus in the stool. Treatment for milk protein intolerance is the same as for milk allergy. Milk protein intolerance is also referred to as milk soy protein intolerance (MSPI).

It is commonplace for milk or milk derivatives to be included in processed foods such as bread, crackers, cookies, cakes, prepared meats, "soy cheese", soups, gravies, crisps, margarine, and products labelled "non-dairy", such as whipped topping and creamer (non-dairy simply means less than 0.5% milk by weight).

In some cases, heating the dairy product to force an exothermic chemical reaction can denature the proteins, (i.e. baking bread, or other baked goods). Only the ingredients that are chemically reacting will denature.

Also, many processed foods that do not contain milk may be processed on equipment contaminated with dairy foods, which may cause an allergic reaction in some sensitive individuals.

Milk Avoidance and Replacement for Infants

Since milk protein may be transferred from a breastfeeding mother to an allergic infant, lactating mothers are put on a dairy elimination diet. For formula fed infants, milk substitute formulas are used to provide a complete source of nutrition. Milk substitutes include soy based formulas, hypoallergenic formulas based on partially or extensively hydrolyzed protein, and free amino acid-based formulas.

Non-milk derived amino acid-based formulas, known as amino acid formulas or elemental formulas, are considered the gold standard in the treatment of cows milk allergy.

Hydrolyzed formulas come in partially hydrolyzed and extensively hydrolyzed varieties. Partially hydrolyzed formulas (PHFs) are characterized by a larger proportion of long chain peptides and are considered more palatable. However, they are intended for milder cases and are not considered suitable for treatment of moderate to severe milk allergy or intolerance.

Extensively hydrolyzed formulas (EHFs) are composed of proteins that have been largely broken down into free amino acids and short peptides. Casein and whey are the most commonly used sources of protein in hydrolyzed formulas because of their high nutritional quality and their amino acid composition.

Soy based formula may or may not pose a risk of allergic sensitivity, as some infants who are allergic to milk may also be allergic to soy. Also soy based formula are not recommended for infants under 6 months. However, for infants with multiple allergies there are rice milk or oat milk based formulas available.

Milk Substitution for Children and Adults

There are many commercially available replacements for milk for children and adults - Rice milk, soy milk, oat milk, coconut milk and almond milk are also sometimes used as milk substitutes, but are not suitable nutrition for infants. However, special infant formula based on soy, rice, almonds or carob seeds is commercially available. Fruit juices supplemented with calcium which may provide an alternative for adults and children. If on an avoidance diet, it is important that dietary advice is taken as a replacement source of calcium may need to be found to prevent the longer term risk of calcium deficiency and osteoporosis. Very good dietary sources of calcium are for example sesame seeds, hemp seeds and some kinds of tofu.

Accidental Exposure

Treatment for accidental ingestion of milk products by allergic individuals varies depending on the sensitivity of the allergic person. Frequently medications such as an Epinephrine pen or an Antihistamine such as Diphenhydramine (Benadryl) are prescribed by an allergist in case of accidental ingestion. Milk allergy can cause anaphylaxis, a severe, life threatening allergic reaction.

Milk allergies are common in infants but are usually outgrown within the first 2–3 years of life.

Statistics

Milk allergy is the most common food allergy in early childhood. It affects somewhere between 2% and 3% of infants in developed countries, but approximately 85–90% of affected children lose clinical reactivity to milk once they surpass 3 years of age.

Between 13% and 20% of children allergic to milk are also allergic to beef.

Pasteurization and Sterilization of Milk

Pasteurization is used to kill harmful microorganisms by heating the milk for a short time and then cooling it for storage and transportation. Pasteurized milk still is perishable, however, and must be stored cold by both suppliers and consumers. Dairies print expiration dates on each container, after which stores will remove any unsold milk from their shelves. The process destroys the vitamin C content of the raw milk.

A newer process, ultrapasteurization or ultra-high temperature treatment (UHT), heats the milk to a higher temperature for a shorter amount of time. This extends its shelf life and allows the milk to be stored unrefrigerated because of the longer lasting sterilization effect, but it affects the taste adversely.

Liquid milk for consumption is mostly either pasteurized or sterilized. Pasteurization is a mild process, designed to inactivate the major pathogenic and spoilage bacteria in raw milk. Further improvements in shelf-life can be obtained by careful control of post-pasteurization contamination (PPC), by use of good quality raw milk and manipulations of the processing conditions. Eventually, however, such milks will spoil due to survival and growth of thermoduric bacteria or any post-pasteurization contaminants.

To keep milk for longer than few days at ambient temperature, it needs to be sterilized. The traditional process involves heating milk in a sealed container in the temperature range 114-120 Celcius degree for 20-30 minutes. More recently UHT processes have been introduced. These are continuous sterilization processes and involve temperatures in excess of 135 Celcius degree for times of greater than 1s, followed by aseptic packaging.

One of the main purposes of heat treatment is to reduce the microbial population in raw milk. Also, when milk is heated enzymes are inactivated, chemical reactions take place and there are changes in physical properties. Some important ones are a decrease in pH, precipitation of calcium phosphate, denaturation of whey proteins and interaction with casein, Maillard browning and modifications to the casein micelle.

The two most important kinetic parameters are the rate of reaction or inactivation at a constant temperature and the effect of temperature change on reaction rate. The heat resistance of vegetative bacteria and microbial spores at a constant temperatures is characterized by their decimal reduction time (D value), this is the time required to reduce the population of 90% or one decimal reduction (one log cycle). The number of decimal reductions (log N0/N) can be evaluated from:

log (N0/N) = heating time/D

where N0 = the initial population, N = final population

The two important points follow from this. Firstly, it is not possible to achieve 100% reduction. Secondly, for a specified heat treatment, the final population will increase as the initial population increases.

Pasteurization

The first stage in the history of pasteurization between 1857 and the end of the nineteenth century might well be called the medical stage, as the main history in heat-treating milk came chiefly from the medical profession interested in infant feeding. In 1927, North and Park established a wide range of temperature-time conditions to inactivate tubercle bacillus.

HTST (high temperature-short time) continuous processes were developed between 1920 and 1927 and for some time the ability of the HSTS process to produce safe milk was questioned. One method of pasteurization produces as good bottle of pasteurized milk as does the other when good methods are used and when conditions are comparable. These included test of the following:

1. Raw milk quality (platform test)
2. Pasteurizability (survival of thermodurics)
3. Efficiency of pasteurization (pathogens and phosphatase)
4. Recontamination (thermophilic and coliform bacteria and the methylene blue test)
5. General bacteria quality, including organisms surviving pasteurization plus contaminating organisms (plate court).

Enzymes in raw milk may give rise to problems in pasteurized milk. However, it is unlikely that bacterial lipases and proteases, which are very heat resistant, will cause problems in pasteurized milks because of their relatively short shelf-life and refrigerated storage conditions.

In general, the lower the storage temperature, the better is the keeping quality. Raw milk is typically stored at 4 Celcius degree, temperatures in the cold chain are slightly higher and they are likely to be higher still in domestic refrigerators. There is a requirement to further increase the shelf-life of pasteurized products, both for the convenience of the consumers and to provide additional protection against temperature abuse. However it is important to avoid the onset of cooked flavour, which would result from more severe pasteurization temperatures.

Sterilization

Sterilization of milk become a commercial proposition in 1894. Milk can be sterilized either in bottles or other sealed containers or by using ultra-high temperature (UHT) processing, which involves continuous sterilization followed by aseptic packaging.

Foods have been sterilized in sealed containers, such as cans, for over 200 years. Milk was originally sterilized in glass bottles sealed with a crown cork but more recently plastic bottles are used. The main aim is to inactivate heat-resistant spores, thereby producing commercially sterile product with an extended shelf-life.

Ultra-high temperature (UHT) offers some distinct advantages over in-container sterilization. Chemical reactions are less temperature sensitive so the use of higher temperatures, combined with more rapid heating and cooling rates, helps to reduce the amount of chemical reaction. There is also a choice of indirect heat exchangers for milk, such as plate or tubular types, as well as direct steam injection or infusion plants, all of which heat products at different rates and shear conditions.

For extended shelf-life and UHT products, aseptic packaging should be used of which a number are available. They are involve putting a sterile product into a sterile container in an aseptic environment. Superheated steam has been used for sterilization of cans. Irradiation may be used for plastic bags.

Package should be inspected regularly to ensure that they are air-tight, again focusing upon those more critical part of the process. Sterilization procedures should be verified. The seal integrity of the package should be monitored as well as the overall microbial quality of packaging material itself. Rinsing, cleaning, and disinfecting procedures are also very important.

Concentrated Milks

Concentrated milk is the milk in which solids have been concentrated by removal of water. Water may be removed by evaporation, by sublimation or by partial freezing followed by removal of ice crystals.

The first method is the most common. The higher concentration of solids which results, provides the product some protection against spoilage by certain chemical actions or microbial agents.

The degree of this protection is closely related to the extent of concentration; the greater the proportion of the water removed, the more the protection.

Whole milk, skim, or standardized milk are used in the preparation of concentrated milks. Partially condensed milk or skim, sweetened or unsweetened condensed milk or skim, evaporated milk or skim, and khoa are concentrated milks. Prevention of Food Adulteration (P.F.A) Act, Government of India uses the term "evaporated" to I4er to unsweetened condensed milk or skim.

Dried Milks and Dry Whey Products

Powdered milk is a manufactured dairy product made by evaporating milk to dryness. One purpose of drying milk is to preserve it; milk powder has a far longer shelf life than liquid milk and does not need to be refrigerated, due to its low moisture content. Another purpose is to reduce its bulk for economy of transportation. Powdered milk and dairy products include such items as dry whole milk, non-fat dry milk, dry buttermilk, dry whey products and dry dairy blends. Many dairy products exported conform to standards laid out in Codex Alimentarius.

Powdered milk is used for food and health (nutrition), and atypically also in biotechnology (saturating agent).

History and Manufacture

While Marco Polo wrote of Mongolian Tatar troops in the time of Kublai Khan carrying sun-dried skimmed milk as "a kind of paste," the first usable commercial production process for dried milk was invented by the Russian chemist M. Dirchoff in 1832. In 1855, T.S. Grimwade took a patent on a dried milk procedure, though a William Newton had patented a vacuum drying process as early as 1837.

In modern times, powdered milk is usually made by spray drying nonfat skim milk, whole milk, buttermilk or whey. Pasteurized milk is first concentrated in an evaporator to approximately 50% milk solids. The resulting concentrated milk is then sprayed into a heated chamber where the water almost instantly evaporates, leaving fine particles of powdered milk solids.

Alternatively, the milk can be dried by drum drying. Milk is applied as a thin film to the surface of a heated drum, and the dried milk solids are then scraped off. However, powdered milk made this way tends to have a cooked flavour, due to caramelization caused by greater heat exposure.

Another process is freeze drying, which preserves many nutrients in milk, compared to drum drying.

The drying method and the heat treatment of the milk as it is processed alters the properties of the milk powder, such as its solubility in cold water, its flavour, and its bulk density.

Food and Health Uses

Powdered milk is frequently used in the manufacture of infant formula, confectionery such as chocolate and caramel candy, and in recipes for baked goods where adding liquid milk would render the product too thin. Powdered milk is also widely used in various sweets such as the famous Indian milk balls known as Rasa-Gulla and popular Indian sweet delicacy (sprinkled with desiccated coconut) known as Chum chum (made with skim milk powder).

Powdered milk is also a common item in UN food aid supplies, fallout shelters, warehouses, and wherever fresh milk is not a viable option. It is widely used in many developing countries because of reduced transport and storage costs (reduced bulk and weight, no refrigerated vehicles). As with other dry foods, it is considered nonperishable, and is favoured by survivalists, hikers, and others requiring nonperishable, easy-to-prepare food.

Reconstituting one cup of potable fluid milk from powdered milk requires one cup of potable water and one-third cup of powdered milk.

Whey Products

Whey is the portion of milk left after cheese or casein is made. Whey is about 96% water but contains some valuable protein, much lactose (milk sugar) and various dissolved salts. Years ago, whey was dumped into streams. These streams became polluted as a result; now, whey is processed into a range of dairy products.

Demineralised whey powder is made by removing dissolved salts and water. Whey protein concentrate is made by very fine filtration of whey to concentrate the proteins into a small volume which are then dried to a soluble powder.

Whey protein hydrolysates are made by digesting the protein concentrate with enzymes, and then drying the resulting soup to a powder.

Milk mineral products are rich in calcium, which is extracted from the whey and then dried.

All these products are sold as ingredients to food processing companies who use them to make food products such as custards, confectionery, crab-sticks, sports drinks, baked goods and yoghurts.

Whey is the by-product of cheese and casein manufacture, being what remains of the milk once the cheese or casein is removed. Generally 100 L of milk produces about 12 kg of cheese or about 3 kg of casein. In either case, about 87 L of whey is made as a by-product.

The annual production of whey in New Zealand is about 4 x 109 L

For decades, most of New Zealand.s whey had been used mainly as pig food or run to waste.

It has also been irrigated as liquid fertiliser. Since about 1970, the New Zealand dairy industry has increasingly processed its whey into a range of products in an attempt both to extract greater total value from the parent milk and to minimise the environmental impact of whey discharges. Economic processing of whey has been aided by the progressive amalgamation of small dairy factories into larger ones where much whey arises at one site.

Major Products Made from Whey in New Zealand

Demineralised Whey Powder

Demineralised Whey Powder is produced using similar equipment to that used to produce milk powder. Whey is first subjected to either electrodialysis and/or ion exchange (described below) to reduce the mineral content by up to 90%. The demineralised whey is then evaporated to about 58% total solids and then rapidly cooled to force as much of the lactose as possible into fine crystal form. The cooled slurry is spray dried.

The spray-dried product is about 75% lactose. Spray drying is very fast and gives insufficient time for the lactose to crystallise, so that the lactose dries as a hygroscopic glass. This makes whey powders very sticky under humid conditions. The problem is minimised by first converting as much as possible of the lactose to the á-monohydrate crystal form, which is not hygroscopic. The resulting powder is sold mainly for use in infant formula preparations.

Ion exchange employs tanks full of charged plastic beads called resin. Whey is pumped down through two resin tanks: cation exchange resin in the first tank and an identical tank of anion exchange resin. The cation exchange resins are covered with sulphonate groups which carry a negative charge at all pH values above about pH 2. This resin strips metal ions from the whey and gives up H+ ions in return. The anion exchange resin is covered in quaternary amine groups which maintain a strong positive charge. This resin picks up chloride, sulphate, phosphate, citrate, nitrate and other anions from the whey in return for OH- ions.

When the cation exchange resin becomes saturated, it is regenerated ready for reuse: a solution of H2SO4 is passed through the resin bed so that the H+ ions present in high concentration displace the Ca2+, Mg2+, Na+ and K+ ions extracted from the whey. Similarly, the anion exchange resin is regenerated by NaOH solution.

Electrodialysis is a process employing a stack of porous plastic membranes sandwiched between two electrodes. Whey is pumped through the envelope made by adjacent sheets of membrane. Water is pumped through the flat spaces between those envelopes. The applied voltage encourages cations to move though the porous membranes from the whey into the water towards the cathode. Similarly anions move from whey to water towards the anode. The passage of ions from water back into whey is impeded by the chemistry of the porous membranes used. Each second membrane carries a net negative charge to make it select for the passage of cations. The alternate membranes are anion selective.

Whey Protein Concentrate (WPC)

Whey Protein Concentrate (WPC) is produced using ultrafiltration. This is also a membrane separation, but it selects on the basis of molecular size and is driven by pressure rather than by applied electric field as in the case of electrodialysis. Ultrafiltration retains (in the liquid product termed .retentate.) any insoluble material or solutes larger than about 20 000 Da molecular weight. The rest of the whey stream passes through the membrane, driven by the applied pressure and is called .permeate.. The permeate contains most of the lactose, minerals and water from the whey. The retentate, the volume of which is about 1-4% that of the feed whey, is spray dried to a powder containing 35-85% protein as desired. WPCs are made at low to moderate temperatures so that the proteins remain in their native form and the dried product is highly soluble. New Zealand is now the largest manufacturer of WPC in the world, employing many tens of square kilometres of ultrafiltration membrane. WPCs are used as food product ingredients in hams, custards, confectionery, crab-sticks, cakes, infant formulae, sports drinks and formulated stock foods.

Most WPCs contain 5-7% milkfat in the dry powdered product. This fat originated in the milk and is not removed by the cream separators through which the whey passes before ultrafiltration. The most modern WPCs use either microfiltration (like ultrafiltration but using membranes with pores sized at about 200 000 Da molecular weight) or ion exchange of the proteins themselves, prior to ultrafiltration, to make a protein product almost devoid of all fat. These very high value proteins find favour in clear acid sports beverages such as those for body-builders.

Lactalbumin

Lactalbuminin the traditional name for a product made from the same whey proteins as are in WPCs, but in an insoluble form. Lactalbumin is prepared from whey by heat precipitation under acidic conditions. Here, the major whey proteins denature and aggregate into fine flocculant curd particles. These can be separated from the liquid .serum. by centrifugation, and then washed and dried. Lactalbumin is nutritionally very valuable and is used primarily for baking, for speciality foods and to fortify some pizza cheeses. However, as the protein is insoluble, it is not useful in helping a food to gel or foam or bind together.

Whey Protein Hydrolysates

Whey Protein Hydrolysates are manufactured by enzyme digesting, at controlled temperature and pH, either WPC or lactalbumin raw materials, and then filtering and spray drying the resulting solution. Whey protein hydrolysates find uses in high value specialist nutritional applications such as tube feeding preparations or special dietary supplements. Appropriately hydrolysed proteins lose the ability to induce allergic reactions in susceptible people, and so can be used in hypoallergenic infant formulae. Proteins are chains of amino acids, in which the amine group of one amino acid is bound to the carboxylic acid group of the neighbouring amino acid by an amide bond. Proteolytic enzymes catalyse the hydrolysis of these bonds Small chains of amino acids are called peptides. In a hydrolysate, we might seek to get all the protein into peptides of two to five amino acids, with few free amino acids and no larger peptides. Interestingly, some peptides released from milk proteins can be biologically active.

Some can transport calcium from the gut into the blood during digestion, some can inhibit enzymes in the human body involved in excessive blood pressure and some can induce sleepy feelings. However, a problem is that some peptides can be very bitter on the tongue.

Choice of the right enzymes and careful control of the hydrolysis process are required to make the desired hydrolysate product.

Milk Mineral Products

Milk mineral products rich in natural milk calcium and phosphate are valuable nutritional supplements in today osteoporosis-sensitive world. These products are made by precipitation of the calcium phosphates in whey ultrafiltration permeate under suitable conditions of concentration, pH, time and temperature. The crystals that first precipitate quickly undergo solid state transitions depending on the conditions to which they are subjected. It is necessary to grow calcium phosphate particles to sufficient size to recover them in a good yield by centrifugation and filtration. Milk mineral is used as a natural. Calcium supplement in a growing range of food products including milks, yoghurts, canned milk powders, confectionary and health foods.

Demineralised Permeate Powders

Demineralised Permeate Powders are manufactured in a very similar manner to demineralised whey powders, but use ultrafiltration permeate of whey as a raw material. The spray-dried product contains about 95% lactose.

Alcohol

The lactose in whey can be converted by fermentation by a variety of organisms to products ranging from lactic acid to flavouring materials. Three plants in New Zealand use yeast to ferment the whey ultrafiltration permeate or lactalbumin serum to ethanol. The ethanol is recovered by distillation to yield potable or industrial grade alcohol.

6

Cream Microbiology

Cream is a dairy product that is composed of the higher-butterfat layer skimmed from the top of milk before homogenization. In unhomogenized milk, over time, the lighter fat rises to the top. In the industrial production of cream this process is accelerated by using centrifuges called "separators". In many countries, cream is sold in several grades depending on the total butterfat content. Cream can be dried to a powder for shipment to distant markets.

Cream skimmed from milk may be called "sweet cream" to distinguish it from whey cream skimmed from whey, a by-product of cheese-making. Whey cream has a lower fat content and tastes more salty, tangy and "cheesy".

Cream produced by cows (particularly Jersey cattle) grazing on natural pasture often contains some natural carotenoid pigments derived from the plants they eat; this gives the cream a slight yellow tone, hence the name of the yellowish-white colour, cream. Cream from goat's milk, or from cows fed indoors on grain or grain-based pellets, is white.

Types

Different grades of cream are distinguished by their fat content, whether they have been heat-treated, whipped, etc. In many jurisdictions there are regulations for each type.

United States

In the United States, cream is usually sold as:

- Half and half (10.5–18% fat)
- Light, coffee, or table cream (18–30% fat)

- Medium cream (25% fat)
- Whipping or light Whipping cream (30–36% fat)
- Heavy Whipping cream (36% or more)
- Extra-heavy, double, or manufacturer's cream (38–40% or more), generally not available at retail except at some warehouse and specialty stores.

Not all grades are defined by all jurisdictions, and the exact fat content ranges vary. The above figures are based on the Code of Federal Regulations, Title 21, Part 131 and a small sample of state regulations.

Australia

In Australia, levels of fat in cream are not regulated, therefore labels are only under the control of the manufacturers. A general guideline is as follows:

Extra light (or 'lite'): 12–12.5% fat.

Light (or 'lite'): 18–20% fat.

Pure cream: 35–56% fat, without artificial thickeners.

Thickened cream: 35–36.5% fat, with added gelatine and/or other thickeners to give the cream a creamier texture, also possibly with stabilisers to aid the consistency of whipped cream (this would be the cream to use for whipped cream, not necessarily for cooking)

Single cream: Recipes calling for 'single cream' are referring to pure or thickened cream with about 35% fat.

Double cream: 48–60% fat.

Canada

Canadian cream definitions are similar to those used in the United States, except for that of "light cream." In Canada, "light cream" is low-fat cream, with 5% or 6% fat. Another form of cream available in Canada is "cereal cream", which is approximately mid-way between 5% cream and coffee cream in fat content.

Japan

In Japan, cream sold in supermarkets is usually between 46% and 48% butterfat.

Processing and Additives

Cream may have thickening agents and stabilisers added. Thickeners include sodium alginate, carrageenan, gelatine, sodium bicarbonate, tetrasodium pyrophosphate, and alginic acid). Other processing may be carried out. For example, cream has a tendency to produce oily globules (called "feathering") when added to coffee. The stability of the cream may be increased by increasing the non-fat solids content, which can be done by partial demineralisation and addition of sodium caseinate, although this is expensive.

Other Cream Products

Butter is made by churning cream to separate the butterfat and buttermilk. This can be done by hand or by machine.

Whipped cream is made by whisking or mixing air into cream with more than 30% fat, to turn the liquid cream into a soft solid. Nitrous oxide may also be used to make whipped cream.

Sour cream, common in many countries including the U.S. and Australia, is cream (12 to 16% or more milk fat) that has been subjected to a bacterial culture that produces lactic acid (0.5%+), which sours and thickens it.

Crème fraîche (28% milk fat) is slightly soured with bacterial culture, but not as sour or as thick as sour cream. Mexican crema (or cream espesa) is similar to crème fraîche.

Smetana is a heavy cream product (35-40% milk fat) Central and Eastern European sour cream.

Rjome or rømme is Norwegian sour cream containing 35% milk fat, similar to Icelandic rjómi.

Clotted cream, common in the United Kingdom, is cream that has been slowly heated to dry and thicken it, producing a very high-fat (55%) product. This is similar to Indian malai.

Cream as an Ingredient

Cream is used as an ingredient in many foods, including ice cream, many sauces, soups, stews, puddings, and some custard bases, and is also used for cakes. Irish cream is an alcoholic liqueur which blends cream with whiskey and coffee. Cream is also used in curries such as masala dishes.

Cream (usually light/single cream or half and half) is often added to coffee.

For cooking purposes, both single and double cream can be used in cooking, although the former can separate when heated, usually if there is a high acid content. Most UK chefs always use double cream or full-fat crème fraîche when cream is added to a hot sauce, to prevent any problem with it separating or "splitting". In sweet and savoury custards such as those found in flan fillings, crème brûlées and crème caramels, both types of cream are called for in different recipes depending on how rich a result is called for. It is useful to note that double cream can also be thinned down with water to make an approximation of single cream if necessary.

Other Items Called "Cream"

Many non-edible substances are called creams due merely to their consistency: shoe cream is runny, unlike waxy shoe polish; face cream is a cosmetic. There is generally no restriction on describing non-edible products as creams. Regulations in many jurisdictions restrict the use of the word *cream* for foods. Words such as *creme*, *kreme*, *creame*, or *whipped topping* are often used for products which cannot legally be called cream. In some cases foods can be described as cream although they do not contain predominantly milk fats; for example in Britain "ice cream" does not have to be a dairy product (although it must be labelled "contains non-milk fat"), and salad cream is the customary name for a condiment that has been produced since the 1920s and need contain no cream.

Butter

Butter is a dairy product made by churning fresh or fermented cream or milk. It is generally used as a spread and a condiment, as well as in cooking applications, such as baking, sauce making, and pan frying. Butter consists of butterfat, water and milk proteins.

Most frequently made from cows' milk, butter can also be manufactured from the milk of other mammals, including sheep, goats, buffalo, and yaks. Salt, flavourings and preservatives are sometimes added to butter. Rendering butter produces clarified butter or *ghee*, which is almost entirely butterfat.

Butter is a water-in-oil emulsion resulting from an inversion of the cream, an oil-in-water emulsion; the milk proteins are the emulsifiers. Butter remains a solid when refrigerated, but softens to a spreadable consistency at room temperature, and melts to a thin liquid consistency at 32–35 °C (90–95 °F). The density of butter is 911 g/L (56.9 lb/ft^3).

It generally has a pale yellow colour, but varies from deep yellow to nearly white. Its unmodified colour is dependent on the animals' feed and is commonly manipulated with food colorings in the commercial manufacturing process, most commonly annatto or carotene.

Etymology

The word *butter* derives (via Germanic languages) from the Latin *butyrum*, which is the latinisation of the Greek *bouturon*. This may have been a construction meaning "cow-cheese", from *bous*, "ox, cow" + *turos*, "cheese", but perhaps this is a false etymology of a Scythian word. Nevertheless, the earliest attested form of the second stem, *turos* ("cheese"), is the Mycenaean Greek *tu-ro*, written in Linear B syllabic script. The root word persists in the name butyric acid, a compound found in rancid butter and dairy products such as Parmesan cheese. Another possibility may be an extended derivation from Sanskrit *bhutari*, meaning "the enemy of evil spirits".

In general use, the term "butter" refers to the spread dairy product when unqualified by other descriptors. The word commonly is used to describe puréed vegetable or nut products such as peanut butter and almond butter. It is often applied to spread fruit products such as apple butter. Fats such as cocoa butter and shea butter that remain solid at room temperature are also known as "butters". In addition to the act of applying butter being called "to butter", non-dairy items that have a dairy butter consistency may use "butter' to call that consistency to mind, including food items such as maple butter and Witch's butter and non-food items such as baby bottom butter, hyena butter, and rock butter.

Production

Unhomogenized milk and cream contain butterfat in microscopic globules. These globules are surrounded by membranes made of phospholipids (fatty acid emulsifiers) and proteins, which prevent the fat in milk from pooling together into a single mass. Butter is produced by agitating cream, which damages these membranes and allows the milk fats to conjoin, separating from the other parts of the cream.

Variations in the production method will create butters with different consistencies, mostly due to the butterfat composition in the finished product. Butter contains fat in three separate forms: free butterfat, butterfat crystals, and undamaged fat globules. In the finished product, different proportions of these forms result in different consistencies within the butter; butters with many crystals are harder than butters dominated by free fats.

Churning produces small butter grains floating in the water-based portion of the cream. This watery liquid is called buttermilk—although the buttermilk most common today is instead a directly fermented skimmed milk. The buttermilk is drained off; sometimes more buttermilk is removed by rinsing the grains with water. Then the grains are "worked": pressed and kneaded together. When prepared manually, this is done using wooden boards called scotch hands. This consolidates the butter into a solid mass and breaks up embedded pockets of buttermilk or water into tiny droplets.

Commercial butter is about 80% butterfat and 15% water; traditionally made butter may have as little as 65% fat and 30% water. Butterfat consists of many moderate-sized, saturated hydrocarbon chain fatty acids. It is a triglyceride, an ester derived from glycerol and three fatty acid groups. Butter becomes rancid when these chains break down into smaller components, like butyric acid and diacetly. The density of butter is 0.911 g/cm^3 (0.527 oz/in^3), about the same as ice.

Types

Before modern factory butter making, cream was usually collected from several milkings and was therefore several days old and somewhat fermented by the time it was made into butter. Butter made from a fermented cream is known as cultured butter. During fermentation, the cream naturally sours as bacteria convert milk sugars into lactic acid. The fermentation process produces additional aroma compounds, including diacetly, which makes for a fuller-flavoured and more "buttery" tasting product. Today, cultured butter is usually made from pasteurized cream whose fermentation is produced by the introduction of *Lactococcus* and *Leuconostoc* bacteria. Another method for producing cultured butter, developed in the early 1970s, is to produce butter from fresh cream and then incorporate bacterial cultures and lactic acid.

Using this method, the cultured butter flavour grows as the butter is aged in cold storage. For manufacturers, this method is more efficient, since aging the cream used to make butter takes significantly more space than simply storing the finished butter product. A method to make an artificial simulation of cultured butter is to add lactic acid and flavour compounds directly to the fresh-cream butter; while this more efficient process is claimed to simulate the taste of cultured butter, the product produced is not cultured but is instead flavoured.

Dairy products are often pasteurized during production to kill pathogenic bacteria and other microbes. Butter made from pasteurized fresh cream is called sweet cream butter. Production of sweet cream butter first became common in the 19th century, with the development of refrigeration and the mechanical cream separator. Butter made from fresh or cultured unpasteurized cream is called raw cream butter. Raw cream butter has a "cleaner" cream flavour, without the cooked-milk notes that pasteurization introduces.

Throughout Continental Europe, cultured butter is preferred, while sweet cream butter dominates in the United States and the United Kingdom. Therefore, cultured butter is sometimes labelled *European-style butter* in the United States. Commercial raw cream butter is virtually unheard-of in the United States. Raw cream butter is generally only found made at home by consumers who have purchased raw whole milk directly from dairy farmers, skimmed the cream themselves, and made butter with it. It is rare in Europe as well.

Several spreadable butters have been developed; these remain softer at colder temperatures and are therefore easier to use directly out of refrigeration. Some modify the make-up of the butter's fat through chemical manipulation of the finished product, some through manipulation of the cattle's feed, and some by incorporating vegetable oils into the butter. Whipped butter, another product designed to be more spreadable, is aerated via the incorporation of nitrogen gas—normal air is not used, because doing so would encourage oxidation and rancidity.

All categories of butter are sold in both salted and unsalted forms. Either granular salt or a strong brine are added to salted butter during processing. In addition to enhanced flavour, the addition of salt acts as a preservative.

The amount of butterfat in the finished product is a vital aspect of production. In the United States, products sold as "butter" are required to contain a minimum of 80% butterfat; in practice most American butters contain only slightly more than that, averaging around 81% butterfat. European butters generally have a higher ratio, which may extend up to 85%.

Clarified butter is butter with almost all of its water and milk solids removed, leaving almost-pure butterfat. Clarified butter is made by heating butter to its melting point and then allowing it to cool off; after settling, the remaining components separate by density. At the top, whey proteins form a skin which is removed, and the resulting butterfat is then poured off from the mixture of water and casein proteins that settle to the bottom.

Ghee is clarified butter which is brought to higher temperatures of around 120 °C (250 °F) once the water has cooked off, allowing the milk solids to brown. This process flavours the ghee, and also produces antioxidants which help protect it longer from rancidity. Because of this, ghee can keep for six to eight months under normal conditions.

Cream may be skimmed from whey instead of milk, as a by-product of cheese-making. Whey butter may be made from whey cream. Whey cream and butter have a lower fat content and taste more salty, tangy and "cheesy". They are also cheaper than "sweet" cream and butter.

European Butters

There are several butters produced in Europe with Protected geographical indications, these include:

- Beurre d'Ardenne, from Belgium
- Beurre d'Isigny, from France
- Beurre Charentes-Poitou (Which also includes: Beurre des Charentes and Beurre des Deux-Sèvres under the same classification), from France
- Beurre Rose, from Luxembourg
- Mantequilla de Soria, from Spain
- Mantega de l'Alt Urgell i la Cerdanya, from Spain.

History

The earliest butter would have been from sheep or goat's milk; cattle are not thought to have been domesticated for another thousand years. An ancient method of butter making, still used today in parts of Africa and the Near East, involves a goat skin half filled with milk, and inflated with air before being sealed. The skin is then hung with ropes on a tripod of sticks, and rocked until the movement leads to the formation of butter.

In the Mediterranean climate, unclarified butter spoils quickly—unlike cheese it is not a practical method of preserving the nutrients of milk. The ancient Greeks and Romans seemed to have considered butter a food fit more for the northern barbarians. A play by the Greek comic poet Anaxandrides refers to Thracians as *boutyrophagoi*; "butter-eaters". In *Natural History*, Pliny the Elder calls butter "the most delicate of food among barbarous nations", and goes on to describe its medicinal properties. Later, the physician Galen also described butter as a medicinal agent only.

Historian and linguist Andrew Dalby says that most references to butter in ancient Near Eastern texts should more correctly be translated as ghee. Ghee is mentioned in the Periplus of the Erythraean Sea as a typical trade article around the 1st century CE Arabian Sea, and Roman geographer Strabo describes it as a commodity of Arabia and Sudan. In India, ghee has been a symbol of purity and an offering to the gods—especially Agni, the Hindu god of fire—for more than 3000 years; references to ghee's sacred nature appear numerous times in the Rig Veda, circa 1500–1200 BCE. The tale of the child Krishna stealing butter remains a popular children's story in India today. Since India's prehistory, ghee has been both a staple food and used for ceremonial purposes such as fueling holy lamps and funeral pyres.

Middle Ages

The cooler climates of northern Europe allowed butter to be stored for a longer period before it spoiled. Scandinavia has the oldest tradition in Europe of butter export trade, dating at least to the 12th century. After the fall of Rome and through much of the Middle Ages, butter was a common food across most of Europe, but one with a low reputation, and was consumed principally by peasants. Butter slowly became more accepted by the upper class, notably when the early 16th century Roman Catholic Church allowed its consumption during Lent. Bread and butter became common fare among the middle class and the English, in particular, gained a reputation for their liberal use of melted butter as a sauce with meat and vegetables.

In antiquity, butter was used for fuel in lamps as a substitute for oil. The *Butter Tower* of Rouen Cathedral was erected in the early 16th century when Archbishop Georges d'Amboise authorized the burning of butter instead of oil, which was scarce at the time, during Lent.

Across northern Europe, butter was sometimes treated in a manner unheard-of today: it was packed into barrels (firkins) and buried in peat bogs, perhaps for years. Such "bog butter" would develop a strong flavour as it aged, but remain edible, in large part because of the unique cool, airless, antiseptic and acidic environment of a peat bog. Firkins of such buried butter are a common archaeological find in

Ireland; the Irish National Museum has some containing "a grayish cheese-like substance, partially hardened, not much like butter, and quite free from putrefaction." The practice was most common in Ireland in the 11th–14th centuries; it ended entirely before the 19th century.

Industrialization

Like Ireland, France became well-known for its butter, particularly in Normandy and Brittany. By the 1860s, butter had become so in demand in France that Emperor Napoleon III offered prize money for an inexpensive substitute to supplement France's inadequate butter supplies. A French chemist claimed the prize with the invention of margarine in 1869. The first margarine was beef tallow flavoured with milk and worked like butter; vegetable margarine followed after the development of hydrogenated oils around 1900.

Until the 19th century, the vast majority of butter was made by hand, on farms. The first butter factories appeared in the United States in the early 1860s, after the successful introduction of cheese factories a decade earlier. In the late 1870s, the centrifugal cream separator was introduced, marketed most successfully by Swedish engineer Carl Gustaf Patrik de Laval. This dramatically sped up the butter-making process by eliminating the slow step of letting cream naturally rise to the top of milk. Initially, whole milk was shipped to the butter factories, and the cream separation took place there. Soon, though, cream-separation technology became small and inexpensive enough to introduce an additional efficiency: the separation was accomplished on the farm, and the cream alone shipped to the factory. By 1900, more than half the butter produced in the United States was factory made; Europe followed suit shortly after.

In 1920, Otto Hunziker authored *The Butter Industry, Prepared for Factory, School and Laboratory*, a well-known text in the industry that enjoyed at least three editions (1920, 1927, 1940). As part of the efforts of the American Dairy Science Association, Professor Hunziker and others published articles regarding: causes of tallowiness (an odour defect, distinct from rancidity, a taste defect); mottles (an aesthetic issue related to uneven colour); introduced salts; the impact of creamery metals and liquids; and acidity measurement. These and other ADSA publications helped standardize practices internationally.

Per capita butter consumption declined in most western nations during the 20th century, in large part because of the rising popularity of margarine, which is less expensive and, until recent years, was perceived as being healthier. In the United States, margarine consumption overtook butter during the 1950s and it is still the case today that more margarine than butter is eaten in the U.S. and the EU.

Shape of Butter Sticks

In the United States, butter is usually produced in 4-ounce sticks, wrapped in waxed or foiled paper and sold four to a one-pound carton. This practice is believed to have originated in 1907, when Swift and Company began packaging butter in this manner for mass distribution. Due to historical differences in butter printers (the machines which cut and package butter), these sticks are commonly produced in two different shapes:

- The dominant shape east of the Rocky Mountains is the Elgin, or Eastern-pack shape, named for a dairy in Elgin, Illinois. The sticks are 4^{3}D $_{4}$ inches long and 1 ^{1}D $_{4}$ inches (121 mm × 32 mm) wide and are typically sold stacked two by two in elongated cube-shaped boxes.
- West of the Rocky Mountains, butter printers standardized on a different shape that is now referred to as the Western-pack shape. These butter sticks are 3 ^{1}D $_{8}$ inches long and 1 ^{1}D $_{2}$ inches wide (80 mm × 38 mm) and are usually sold with four sticks packed side-by-side in a flat, rectangular box.

Both sticks contain the same amount of butter, although most butter dishes are designed for Elgin-style butter sticks.

The stick's wrapper is usually marked off as eight tablespoons (120 ml/4.2 imp fl oz; 4.1 US fl oz); the actual volume of one stick is approximately nine tablespoons (130 ml/4.6 imp fl oz; 4.4 US fl oz).

World Wide

In 1997, India produced 1,470,000 metric tons (1,620,000 short tons) of butter, most of which was consumed domestically. Second in production was the United States (522,000 t/575,000 short tons), followed by France (466,000 t/514,000 short tons), Germany (442,000 t/487,000 short tons), and New Zealand (307,000 t/338,000 short tons).

France ranks first in per capita butter consumption with 8 kg per capita per year. In terms of absolute consumption, Germany was second after India, using 578,000 metric tons (637,000 short tons) of butter in 1997, followed by France (528,000 t/582,000 short tons), Russia (514,000 t/567,000 short tons), and the United States (505,000 t/

557,000 short tons). New Zealand, Australia, and the Ukraine are among the few nations that export a significant percentage of the butter they produce.

Different varieties are found around the world. *Smen* is a spiced Moroccan clarified butter, buried in the ground and aged for months or years. Yak butter is important in Tibet; *tsampa*, barley flour mixed with yak butter, is a staple food. Butter tea is consumed in the Himalayan regions of Tibet, Bhutan, Nepal and India. It consists of tea served with intensely flavoured—or "rancid"—yak butter and salt. In African and Asian developing nations, butter is traditionally made from sour milk rather than cream. It can take several hours of churning to produce workable butter grains from fermented milk.

Storage and Cooking

Normal butter softens to a spreadable consistency around 15 °C (60 °F), well above refrigerator temperatures. The "butter compartment" found in many refrigerators may be one of the warmer sections inside, but it still leaves butter quite hard. Until recently, many refrigerators sold in New Zealand featured a "butter conditioner", a compartment kept warmer than the rest of the refrigerator—but still cooler than room temperature—with a small heater. Keeping butter tightly wrapped delays rancidity, which is hastened by exposure to light or air, and also helps prevent it from picking up other odors. Wrapped butter has a shelf life of several months at refrigerator temperatures.

"French butter dishes" or "Acadian butter dishes" involve a lid with a long interior lip, which sits in a container holding a small amount of water. Usually the dish holds just enough water to submerge the interior lip when the dish is closed. Butter is packed into the lid. The water acts as a seal to keep the butter fresh, and also keeps the butter from overheating in hot temperatures. This allows butter to be safely stored on the countertop for several days without spoilage.

Once butter is softened, spices, herbs, or other flavoring agents can be mixed into it, producing what is called a *compound butter* or *composite butter* (sometimes also called *composed butter*). Compound butters can be used as spreads, or cooled, sliced, and placed onto hot food to melt into a sauce. Sweetened compound butters can be served with desserts; such hard sauces are often flavoured with spirits.

Melted butter plays an important role in the preparation of sauces, most obviously in French cuisine. *Beurre noisette* (hazelnut butter) and *Beurre noir* (black butter) are sauces of melted butter cooked until the milk solids and sugars have turned golden or dark brown; they are often finished with an addition of vinegar or lemon juice. Hollandaise and béarnaise sauces are emulsions of egg yolk and melted butter; they are in essence mayonnaises made with butter instead of oil. Hollandaise and béarnaise sauces are stabilized with the powerful emulsifiers in the egg yolks, but butter itself contains enough emulsifiers—mostly remnants of the fat globule membranes—to form a stable emulsion on its own. *Beurre blanc* (white butter) is made by whisking butter into reduced vinegar or wine, forming an emulsion with the texture of thick cream. *Beurre monté* (prepared butter) is melted but still emulsified butter; it lends its name to the practice of "mounting" a sauce with butter: whisking cold butter into any water-based sauce at the end of cooking, giving the sauce a thicker body and a glossy shine—as well as a buttery taste.

In Poland, the butter lamb (*Baranek wielkanocny*) is a traditional addition to the Easter Meal for many Polish Catholics. Butter is shaped into a lamb either by hand or in a lamb-shaped mould.

Butter is also used to make edible decorations to garnish other dishes.

Butter is used for sautéing and frying, although its milk solids brown and burn above 150 °C (250 °F)—a rather low temperature for most applications. The smoke point of butterfat is around 200 °C (400 °F), so clarified butter or ghee is better suited to frying. Ghee has always been a common frying medium in India, where many avoid other animal fats for cultural or religious reasons.

Butter fills several roles in baking, where it is used in a similar manner as other solid fats like lard, suet, or shortening, but has a flavour that may better complement sweet baked goods. Many cookie doughs and some cake batters are leavened, at least in part, by creaming butter and sugar together, which introduces air bubbles into the butter. The tiny bubbles locked within the butter expand in the heat of baking and aerate the cookie or cake. Some cookies like shortbread may have no other source of moisture but the water in the butter. Pastries like pie dough incorporate pieces of solid fat into the dough, which become flat layers of fat when the dough is rolled out. During baking, the fat melts away, leaving a flaky texture. Butter, because of its flavour, is a common choice for the fat in such a dough, but it can be more difficult to work with than shortening because of its low melting point. Pastry makers often chill all their ingredients and utensils while working with a butter dough.

Butter also has many non-culinary, traditional uses which are specific to certain cultures. For instance, in North America, applying butter to the handle of a door is a common prank on April Fools' Day.

Health and Nutrition

Table 1: Butter, Unsalted Nutritional Value Per 100 g (3.5 oz)

Energy	2,999 kJ (717 kcal)
Carbohydrates	0 g
Fat	81 g
Saturated	51 g
Monounsaturated	21 g
Polyunsaturated	3 g
Protein	1 g
Vitamin A equiv.	684 μg (76%)
Vitamin D	60 IU (15%)
Vitamin E	2.32 mg (15%)
Cholesterol	215 mg

Fat percentage can vary.

Percentages are relative to US recommendations for adults.

Source: USDA Nutrient database

According to USDA figures, one tablespoon of butter (14 grams / 0.5 ounces) contains 420 kilojoules (100 kcal), all from fat, 11 grams (0.4 oz) of fat, of which 7 grams (0.25 oz) are saturated fat, and 30 milligrams (0.46 gr) of cholesterol.

Butter consists mostly of saturated fat and is a significant source of cholesterol. For these reasons butter is considered by some to be a contributor to health problems, especially heart disease. Margarine was recommended as a substitute, since it is higher in unsaturated fat and contains little or no cholesterol, but in recent years, it has been shown that the trans fats contained in partially hydrogenated oils used in typical margarines significantly raise undesirable LDL cholesterol levels as well. Trans-fat–free margarines have since been developed.

Proponents of the consumption of organic butter, such as the nutritionist Mary Enig, state that since butter is nutritious and "is rich in short and medium chain fatty acids," this can have a positive effect on health and prevent disease.

Butter contains only traces of lactose, so moderate consumption of butter is not a problem for the lactose intolerant. People with milk allergies need to avoid butter, which contains enough of the allergy-causing proteins to cause reactions.

Butter can perform a useful role in dieting by providing satiety. A small amount added to low fat foods such as vegetables may stave off feelings of hunger.

Icecream and Related Frozen Dairy Desserts

Ice cream is a frozen dessert usually made from dairy products, such as milk and cream, and often combined with fruits or other ingredients and flavours. Most varieties contain sugar, although some are made with other sweeteners. In some cases, artificial flavourings and colorings are used in addition to (or in replacement of) the natural ingredients. This mixture is stirred slowly while cooling to prevent large ice crystals from forming; the result is a smoothly textured ice cream.

The meaning of the term ice cream varies from one country to another. Terms like frozen custard, frozen yogurt, sorbet, gelato and others are used to distinguish different varieties and styles. In some countries, like the USA, the term ice cream applies only to a specific variety, and their governments regulate the commercial use of all these terms based on quantities of ingredients. In others, like Italy and Argentina, one word is used for all the variants. Alternatives made from soy milk, rice milk, and goat milk are available for those who are lactose intolerant or have an allergy to dairy protein, or in the case of soy and rice milk, for those who want to avoid animal products.

History

Precursors of Ice Cream

In the Persian Empire, people would pour grape juice concentrate over snow - in a bowl - and eat this as a treat. In particular this was consumed when the weather was hot. Either snow would be saved in the cool-keeping underground chambers known as "yakhchal" or taken from fresh snow that may still have remained at the top of the mountains by the summer capital - Hagmatana, Ecbatana or Hamedan of today. In 400 BC, the Persians went further and invented a special chilled food, made of rose water and vermicelli which was served to royalty during summers. The ice was mixed with saffron, fruits, and various other flavours.

Ancient civilizations have served ice for cold foods for thousands of years. The BBC reports that a frozen mixture of milk and rice was used in China around 200 BC. The Roman Emperor Nero (37–68) had ice brought from the mountains and combined with fruit toppings. These were some early chilled delicacies.

Arabs were the first to use milk as a major ingredient in its production, sweeten the ice cream with sugar rather than fruit juices, as well as perfect ways for its commercial production. As early as the 10th century, ice cream was widespread amongst many of the Arab world's major cities, such as Baghdad, Damascus and Cairo. Their version of ice cream was produced from milk or cream and often some yoghurt similar to Ancient Greek recipes, flavoured with rosewater as well as dried fruits and nuts. It is believed that this was based on older Ancient Arab, Mesopotamian, Greek or Roman recipes, which were probably the first and precursors to Persian faloodeh. In 62 AD, the Roman emperor Nero sent slaves to the Apennine mountains to collect snow to be flavoured with honey and nuts.

Maguelonne Toussaint-Samat asserts in her *History of Food*, "the Chinese may be credited with inventing a device to make sorbets and ice cream. They poured a mixture of snow and saltpetre over the exteriors of containers filled with syrup, for, in the same way as salt raises the boiling-point of water, it lowers the freezing-point to below zero." (Toussaint does not provide historical documentation for this.) Some distorted accounts claim that in the age of Emperor Yingzong, Song Dynasty (960-1279) of China, was written by the poet Yang Wanli. Actually, and has nothing to do with ice cream. It has also been claimed that, in the Yuan Dynasty, Kublai Khan enjoyed ice cream and kept it a royal secret until Marco Polo visited China and took the technique of making ice cream to Italy. Others have argued that the Chinese didn't drink milk during that period, whereas the Italians had arguably been making something resembling ice cream before Marco Polo returned to Italy. In any case, no known ice cream recipes appear to stem from ancient Chinese sources.

In the sixteenth century, the Mughal emperors used relays of horsemen to bring ice from the Hindu Kush to Delhi, where it was used in fruit sorbets.

When Italian duchess Catherine de' Medici married the duc d'Orléans in 1533, she is said to have brought with her Italian chefs who had recipes for flavoured ices or sorbets, and introduced them in France. One hundred years later, Charles I of England was supposedly so impressed by the "frozen snow", he offered his own ice cream maker a lifetime pension in return for keeping the formula secret, so ice cream could be a royal prerogative. There is no historical evidence to support these legends, which first appeared during the 19th century.

The first recipe for flavoured ices in French appears in 1674, in Nicholas Lemery's *Recueil de curiositéz rares et nouvelles de plus admirables effets de la nature*. Recipes for *sorbetti* saw publication in the 1694 edition of Antonio Latini's *Lo Scalco alla Moderna* (The Modern Steward). Recipes for flavoured ices begin to appear in François Massialot's *Nouvelle Instruction pour les Confitures, les Liqueurs, et les Fruits* starting with the 1692 edition. Massialot's recipes result in a coarse, pebbly texture. Latini claims that the results of his recipes should have the fine consistency of sugar and snow.

True Ice Cream

Ice cream recipes first appear in 18th century England and America. A recipe for ice cream was published in *Mrs. Mary Eales's Receipts* in London 1718.

To ice CREAM. Take Tin Ice-Pots, fill them with any Sort of Cream you like, either plain or sweeten'd, or Fruit in it; shut your Pots very close; to six Pots you must allow eighteen or twenty Pound of Ice, breaking the Ice very small; there will be some great Pieces, which lay at the Bottom and Top: You must have a Pail, and lay some Straw at the Bottom; then lay in your Ice, and put in amongst it a Pound of Bay-Salt; set in your Pots of Cream, and lay Ice and Salt between every Pot, that they may not touch; but the Ice must lie round them on every Side; lay a good deal of Ice on the Top, cover the Pail with Straw, set it in a Cellar where no Sun or Light comes, it will be froze in four Hours, but it may stand longer; than take it out just as you use it; hold it in your Hand and it will slip out. When you would freeze any Sort of Fruit, either Cherries, Raspberries, Currants, or Strawberries, fill your Tin-Pots with the Fruit, but as hollow as you can; put to them Lemmonade, made with Spring-Water and Lemmon-Juice sweeten'd; put enough in the Pots to make the Fruit hang together, and put them in Ice as you do Cream.

The earliest reference to ice cream given by the *Oxford English Dictionary* is from 1744, reprinted in a magazine in 1877. *1744 in Pennsylvania Mag. Hist. & Biogr. (1877) I. 126 Among the rarities..was some fine ice cream, which, with the strawberries and milk, eat most deliciously*. The 1751 edition of *The Art of Cookery made Plain and Easy* by Hannah Glasse features a recipe for ice cream. OED gives her recipe: *H. GLASSE Art of Cookery (ed. 4) 333 (heading) To make Ice Cream..set it [sc. the cream] into the larger Bason. Fill it with Ice, and a Handful of Salt.*

1768 saw the publication of *L'Art de Bien Faire les Glaces d'Office* by M. Emy, a cookbook devoted entirely to recipes for flavoured ices and ice cream.

Ice cream was introduced to the United States by Quaker colonists who brought their ice cream recipes with them. Confectioners sold ice cream at their shops in New York and other cities during the colonial era. Ben Franklin, George Washington, and Thomas Jefferson were known to have regularly eaten and served ice cream. First Lady Dolley Madison is also closely associated with the early history of ice cream in the United States. One respected history of ice cream states that, as the wife of U.S. President James Madison, she served ice cream at her husband's Inaugural Ball in 1813.

Around 1832, Augustus Jackson, an African American confectioner, not only created multiple ice cream recipes, but he also invented a superior technique to manufacture ice cream.

In 1843, Nancy Johnson of Philadelphia was issued the first U.S. patent for a small-scale handcranked ice cream freezer. The invention of the ice cream soda gave Americans a new treat, adding to ice cream's popularity. This cold treat was probably invented by Robert Green in 1874, although there is no conclusive evidence to prove his claim.

The ice cream sundae originated in the late 19th century. Several men claimed to have created the first sundae, but there is no conclusive evidence to back up any of their stories. Some sources say that the sundae was invented to circumvent blue laws, which forbade serving sodas on Sunday. Towns claiming to be the birthplace of the sundae include Buffalo, New York; Two Rivers, Wisconsin; Ithaca, New York; and Evanston, Illinois. Both the ice cream cone and banana split became popular in the early 20th century. Several food vendors claimed to have invented the ice cream cone at the 1904 World's Fair in St.

Louis, MO. Europeans were eating cones long before 1904. In the UK, ice cream remained an expensive and rare treat, until large quantities of ice began to be imported from Norway and the US in the mid Victorian era. A Swiss-Italian businessman, Carlo Gatti, opened the first ice cream stall outside Charing Cross station in 1851, selling scoops of ice cream in shells for one penny.

The history of ice cream in the 20th century is one of great change and increases in availability and popularity. In the United States in the early 20th century, the ice cream soda was a popular treat at the soda shop, the soda fountain, and the ice cream parlor. During American Prohibition, the soda fountain to some extent replaced the outlawed alcohol establishments such as bars and saloons.

Ice cream became popular throughout the world in the second half of the 20th century after cheap refrigeration became common. There was an explosion of ice cream stores and of flavours and types. Vendors often competed on the basis of variety. Howard Johnson's restaurants advertised "a world of 28 flavours." Baskin-Robbins made its 31 flavours ("one for every day of the month") the cornerstone of its marketing strategy. The company now boasts that it has developed over 1000 varieties. One important development in the 20th century was the introduction of soft ice cream. A chemical research team in Britain (of which a young Margaret Thatcher was a member) discovered a method of doubling the amount of air in ice cream, which allowed manufacturers to use less of the actual ingredients, thereby reducing costs. It made possible the soft ice cream machine in which a cone is filled beneath a spigot on order. In the United States, Dairy Queen, Carvel, and Tastee-Freez pioneered in establishing chains of soft-serve ice cream outlets.

Technological innovations such as these have introduced various food additives into ice cream, notably the stabilizing agent gluten, to which some people have an intolerance. Recent awareness of this issue has prompted a number of manufacturers to start producing gluten-free ice cream. The 1980s saw a return of the older, thicker ice creams being sold as "premium" and "superpremium" varieties under brands such as Ben & Jerry's and Häagen-Dazs.

Production

Before the development of modern refrigeration, ice cream was a luxury reserved for special occasions. Making it was quite laborious; ice was cut from lakes and ponds during the winter and stored in holes in the ground, or in wood-frame or brick ice houses, insulated by straw. Many farmers and plantation owners, including U.S. Presidents George Washington and Thomas Jefferson, cut and stored ice in the winter for use in the summer. Frederic Tudor of Boston turned ice harvesting and shipping into a big business, cutting ice in New England and shipping it around the world.

Ice cream was made by hand in a large bowl placed inside a tub filled with ice and salt. This was called the pot-freezer method. French confectioners refined the pot-freezer method, making ice cream in a sorbetière (a covered pail with a handle attached to the lid). In the pot-freezer method, the temperature of the ingredients is reduced by the mixture of crushed ice and salt. The salt water is cooled by the ice, and the action of the salt on the ice causes it to (partially) melt, absorbing latent heat and bringing the mixture below the freezing point of pure water. The immersed container can also make better thermal contact with the salty water and ice mixture than it could with ice alone.

The hand-cranked churn, which also uses ice and salt for cooling, replaced the pot-freezer method. The exact origin of the hand-cranked freezer is unknown, but the first U.S. patent for one was #3254 issued to Nancy Johnson on September 9, 1843. The hand-cranked churn produced smoother ice cream than the pot freezer and did it quicker. Many inventors patented improvements on Johnson's design.

In Europe and early America, ice cream was made and sold by small businesses, mostly confectioners and caterers. Jacob Fussell of Baltimore, Maryland was the first to manufacture ice cream on a large scale. Fussell bought fresh dairy products from farmers in York County, Pennsylvania, and sold them in Baltimore. An unstable demand for his dairy products often left him with a surplus of cream, which he made into ice cream. He built his first ice-cream factory in Seven Valleys, Pennsylvania, in 1851. Two years later, he moved his factory to Baltimore. Later, he opened factories in several other cities and taught the business to others, who operated their own plants. Mass production reduced the cost of ice cream and added to its popularity.

The development of industrial refrigeration by German engineer Carl von Linde during the 1870s eliminated the need to cut and store natural ice and when the continuous-process freezer was perfected in 1926, it allowed commercial mass production of ice cream and the birth of the modern ice cream industry.

The most common method for producing ice cream at home is to use an ice cream maker, in modern times generally an electrical device that churns the ice cream mixture while cooled inside a household freezer, or using a solution of pre-frozen salt and water, which gradually melts while the ice cream freezes. Some more expensive models have an inbuilt freezing element. A newer method of making home-made ice cream is to add liquid nitrogen to the mixture while stirring it using a spoon or spatula. Some ice cream recipes call for making a custard, folding in whipped cream, and immediately freezing the mixture.

Commercial Delivery

Ice cream can be mass-produced and thus is widely available in developed parts of the world. Ice cream can be purchased in large cartons (vats and squrounds) from supermarkets and grocery stores, in smaller quantities from ice cream shops, convenience stores, and milk bars, and in individual servings from small carts or vans at public events. In Turkey and Australia, ice cream is sometimes sold to beach-goers from small powerboats equipped with chest freezers. Some ice cream distributors sell ice cream products from traveling refrigerated vans or carts (commonly referred to in the US as "ice cream trucks"), sometimes equipped with speakers playing children's music. Traditionally, ice cream vans in the United Kingdom make a music box noise rather than actual music.

Dietary

Ice cream may have the following composition:

- greater than 10% milkfat and usually between 10% and as high as 16% fat in some premium ice creams
- 9 to 12% milk solids-not-fat: this component, also known as the serum solids, contains the proteins (caseins and whey proteins) and carbohydrates (lactose) found in milk
- 12 to 16% sweeteners: usually a combination of sucrose and glucose-based corn syrup sweeteners
- 0.2 to 0.5% stabilisers and emulsifiers
- 55% to 64% water which comes from the milk or other ingredients.

These compositions are percentage by weight. Since ice cream can contain as much as half air by volume, these numbers may be reduced by as much as half if cited by volume. In terms of dietary considerations, The percentages by weight are more relevant. Even the low fat products have high caloric content: Ben and Jerry's No Fat Vanilla Fudge contains 150 calories per half cup due to its high sugar content.

Ice Cream around the World

Ice Cream Cone

Mrs Marshall's Cookery Book, published in 1888, endorsed serving ice cream in cones, but the idea definitely predated that. Agnes Marshall was a celebrated cookery writer of her day and helped to popularise ice cream. She patented and manufactured an ice cream maker and was the first person to suggest using liquefied gases to freeze ice cream after seeing a demonstration at the Royal Institution.

Reliable evidence proves that ice cream cones were served in the 19th century, and their popularity increased greatly during the St. Louis World's Fair in 1904. According to legend, at the World's Fair an ice cream seller had run out of the cardboard dishes used to put ice cream scoops in, so they could not sell any more produce.

Next door to the ice cream booth was a Syrian waffle booth, unsuccessful due to intense heat; the waffle maker offered to make cones by rolling up his waffles and the new product sold well, and was widely copied by other vendors.

Other Frozen Desserts

The following is a partial list of ice cream-like frozen desserts and snacks:

- Ais kacang: a dessert in Malaysia and Singapore made from shaved ice, syrup, and boiled red bean and topped with evaporated milk. Sometimes, other small ingredients like raspberries and durians are added in too.
- Dondurma: Turkish ice cream, made of salep and mastic resin
- Frozen custard: at least 10% milk fat and at least 1.4% egg yolk and much less air beaten into it, similar to Gelato, fairly rare. Known in Italy as Semifreddo.
- Frozen yogurt: a low fat or fat free alternative made with yogurt
- Gelato: an Italian frozen dessert having a lower milk fat content than ice cream and stabilised with ingredients such as eggs.
- Halo-halo: a popular Filipino dessert that is a mixture of shaved ice and milk to which are added various boiled sweet beans and fruits, and served cold in a tall glass or bowl.

- Ice milk: less than 10% milk fat and lower sweetening content, once marketed as "ice milk" but now sold as *low-fat ice cream* in the United States.
- Ice pop (or lolly): frozen fruit puree, fruit juice, or flavoured sugar water on a stick or in a flexible plastic sleeve.
- Kulfi: Believed to have been introduced to South Asia by the Mughal conquest in the 16th century; its origins trace back to the cold snacks and desserts of Arab and Mediterranean cultures.
- Mellorine: non-dairy, with vegetable fat substituted for milk fat
- Parevine: Kosher non-dairy frozen dessert established in 1969 in New York
- Sherbet: 1–2% milk fat and sweeter than ice cream.
- Sorbet: fruit puree with no dairy products
- Snow cones, made from balls of crushed ice topped with sweet syrup served in a paper cone, are consumed in many parts of the world. The most common places to find snow cones in the United States are at amusement parks.
- Maple toffee: A popular springtime treat in maple-growing areas is maple toffee, where maple syrup boiled to a concentrated state is poured over fresh snow congealing in a toffee-like mass, and then eaten from a wooden stick used to pick it up from the snow.

Using Liquid Nitrogen

Using liquid nitrogen to freeze ice cream is an old idea and has been used for many years to harden ice cream. The use of liquid nitrogen in the primary freezing of ice cream, that is to effect the transition from the liquid to the frozen state without the use of a conventional ice cream freezer, has only recently started to see commercialization. Some commercial innovations have been documented in the National Cryogenic Society Magazine "Cold Facts". The most noted brands are Dippin' Dots, Blue Sky Creamery, Project Creamery, and Sub Zero Cryo Creamery. The preparation results in a column of white condensed water vapor cloud, reminiscent of popular depictions of witches' cauldrons. The ice cream, dangerous to eat while still "steaming," is allowed to rest until the liquid nitrogen is completely vaporised. Sometimes ice cream is frozen to the sides of the container, and must be allowed to thaw.

Making ice cream with liquid nitrogen has advantages over conventional freezing. Due to the rapid freezing, the crystal grains are smaller, giving the ice cream a creamier texture, and allowing one to get the same texture by using less milkfat. Such ice crystals will grow very quickly via the processes of recrystallization thus obviating the original benefits unless steps are taken to inhibit ice crystal growth.

Gelato

Gelato is Italy's regional variant of ice cream. As such, gelato is made with some of the same ingredients as most other frozen dairy desserts. Milk, cream, various sugars, flavoring including fruit and nut purees are the main ingredients.

Gelato is different from some other ice creams because it has a lower butterfat content. Gelato typically contains 4-8% butterfat, versus 14% for many other ice creams. Gelato generally has slightly lower sugar content, averaging between 16-22% versus approximately 21% for most ice creams. Non-fat milk is added as a solid. The sugar content in gelato is precisely balanced with the water content to act as an anti-freeze to prevent the gelato from freezing solid. Types of sugar used include sucrose, dextrose, and invert sugar to control apparent sweetness. Typically, gelato and Italian sorbet contain a stabilizing base. Egg yolks are used in yellow custard-based gelato flavours, including zabaione and creme caramel.

The mixture for gelato is typically made using a hot process, which includes pasteurization. White base is heated to 85°C (185°F). Heating the mix to 90°C (194°F) is essential for chocolate gelato, which is traditionally flavoured with cocoa powder. Yellow custard base, which contains egg yolks, is heated to 65°C (149°F). The gelato mix must age for several hours after pasteurization is complete for the milk proteins to hydrate, or bind, with water in the mix. This hydration reduces the size of the ice crystals, making a smoother texture in the final product. A non-traditional cold mix process is popular among some gelato makers in the United States.

Unlike most commercial ice creams in the United States, which are frozen with a continuous assembly line freezer, gelato is frozen very quickly in individual small batches in a batch freezer. The batch freezer incorporates air or overage into the mix as it freezes. Unlike most American-style ice creams, which can have an overage of up to 50%, gelato generally has between 20% and 35% overage. This results in a denser product with more intense flavour than many U.S. style ice creams. U.S. style ice cream, with a higher fat content, can be stored in a freezer for months. High-quality artisan gelato holds its peak flavour and texture (from delicate ice crystals) only for several days, even when stored carefully at the proper temperature. This is why *gelaterias* typically make their own gelato on the premises or nearby.

History

The history of gelato dates back to frozen desserts served in ancient Rome and Egypt made from ice and snow brought down from mountaintops and preserved below ground. Later, gelato appeared during banquets at the Medici court in Florence. In fact, the Florentine cook Bernardo Buontalenti is said to have invented modern ice cream in 1565, as he presented his recipe and his innovative refrigerating techniques to Catherine de' Medici. She in turn brought the novelty to France, where in 1686 the Sicilian fisherman Francesco Procopio dei Coltelli perfected the first ice cream machine. The popularity of gelato among larger shares of the population however only increased in the 1920s-1930s as in the northern Italian city of Varese, where the first mobile gelato cart was developed.

Gelato is typically flavoured with fresh fruit purees, cocoa and/or nut pastes. If other ingredients such as chocolate flakes, nuts, small confections, cookies, or biscuits are added, they are added after the gelato is frozen. Gelato made with fresh fruit, sugar, and water and without dairy ingredients is sorbet. Use of the word sorbetto is a common affectation in the United States. In Italy, the word sorbetto refers to a cocktail.

Stracciatella

Stracciatella (from Italian *stracciato*, "torn apart") is an Italian egg-drop soup usually said to be "alla Romana" ("the way it's done in Rome"), but also popular in Marche and Emilia Romagna. It is prepared by beating eggs and adding grated parmesan cheese, salt, pepper, nutmeg, and sometimes semolina, and then adding this mixture to boiling broth. The broth is set whirling first with a whisk, and the beaten egg mixture added in a slow stream to produce the *stracciatelle* ("little shreds") of cooked egg in the broth, which is clarified by the process.

In Gelato

In Italy and Germany, gelato with a vanilla base and chocolate shavings is also called stracciatella or stracciatella ice. It is somewhat analogous to chocolate chip ice cream in North America though the chocolate is intended to be less chunky and more integrated with the gelato.

In Cheese

A particular kind of mozzarella (soft cheese) is also called stracciatella. Stracciatella is used as stuffing for the burrata from the Murgia region in Puglia. It is made with torn pieces of mozzarella and cream.

Frozen Custard

Frozen custard is a cold dessert similar to ice cream, made with eggs in addition to cream and sugar.

In the United States the Food and Drug Administration requires products marketed as frozen custard to contain at least 10 percent milkfat and 1.4 percent egg yolk solids. If it has less egg yolk solids, it is considered ice cream. In the United Kingdom frozen custard is not differentiated from other frozen desserts. Instead, if given a name other than frozen desserts they may be referred to as ice creams.

History

Most early ice cream recipes would be called frozen custard today, because of naming conventions. The modern term of frozen custard is primarily used as a method of brand differentiation and to comply with FDA requirements. One claim traces the invention of the term frozen custard to Coney Island in 1919. However, recipes for custard based ices existed before then, especially in France. In some areas of the United States, "frozen custard" or "custard" has become a synonym for soft serve.

Home Creation

The vast majority of ice cream desserts created at home are frozen custards. Despite commercialization of a limited number of flavours, home cooks can make frozen custard into innumerable flavours. Extra egg-yolk and a brief heating is used in order to make a custard base for the ice. This replaces commercially used ice cream additives such as gluten in order to create the same consistency. It is possible that the heating of the cream containing milk was initially done as a form of pasteurization, with the side effect of creating a custard. At home, frozen custard can be made with a standard ice cream maker or with the freeze and scrape method.

Commercial Creation

Using a process called overrun, air is blended into the mixture of ingredients until its volume increases by approximately 20%. By comparison, ice cream may have an overrun as large as 100%, meaning half of the final product is composed of air. The high percentage of butterfat and egg yolk gives frozen custard a thick, creamy texture and a smoother consistency than ice cream. Frozen custard can be served at –8°C (18°F), warmer than the –12°C (10°F) at which ice cream is served, in order to make a soft serve product.

Another difference between commercially produced frozen custard and commercial ice cream is the way the custard is frozen. The mix enters a refrigerated tube and, as it freezes, blades scrape the product cream off the barrel walls. The now frozen custard is discharged directly into containers from which it can be served. The speed with which the product leaves the barrel minimizes the amount of air in the product but more importantly ensures that the ice crystals formed are very small.

Frozen custard is usually prepared fresh at the place of sale, rather than stored; however, it is occasionally available in supermarkets or by mail order. Generally, modern frozen custard stands provide only three different flavours per day: vanilla, chocolate, and a unique "flavour of the day." The older vintage custard stands tend to have a dozen or so standard flavours. Some locations try to emulate the Coney Island feel (sometimes cited as the birthplace of the term frozen custard) by serving other diner foods along with their premium dessert.

Semifreddo, a Class of Semi-frozen Desserts

Semifreddo (Italian: "half cold") is a class of semi-frozen desserts, typically ice-cream cakes, semi-frozen custards, and certain fruit tarts. It has the texture of frozen mousse because it is usually produced by uniting two equal parts of ice cream and whipped cream. Such a dessert's Spanish counterpart is called semifrío. In Italian cuisine, the Semifreddo is commonly made with gelato as a primary ingredient. It is typical of the Italian region of Emilia-Romagna.

Sorbet

Sorbet is a frozen dessert made from sweetened water flavoured with fruit (typically juice or puree), wine, and/or liqueur. The origin of sorbet is variously explained as either a Roman invention, or a Middle Eastern drink *charbet*, made of sweetened fruit juice and water. The term *sherbet* or *charbet* is derived from Turkish: *þerbat/* °erbet, "sorbet", from the Persian *sharbat*. Sorbet is sometimes served between courses as a way to cleanse the palate before the main course.

Classification and Description

Sorbet is often confused with Italian ice and often taken to be the same as sherbet.

Sorbets/sherbets may also contain alcohol, which lowers the cold temperature, resulting in softer texture. In the UK, *sherbet* refers to a fizzy powder, and only the term *sorbet* would be used.

Whereas ice cream is based on dairy products with air copiously whipped in, sorbet has neither, which makes for a dense and extremely flavorful product. Sorbet is served as a non-fat or low-fat alternative to ice cream.

In Italy a similar though crunchier textured dish called granita is made. As the liquid in granita freezes it forms noticeably large-size crystals, which are left unstirred. Granita is also often sharded with a fork to give an even crunchier texture when served.

Agraz is a type of sorbet, usually associated with the Maghreb and north Africa. It is made from almonds, verjuice, and sugar. It has a strongly acidic flavour, because of the verjuice. (*Larousse Gastronomique*)

Early History and Folklore

One account says that Marco Polo brought a recipe for a sorbet-like dessert on his way back to Italy from China in the late 13th century, as written in an account of his journey, *The Travels of Marco Polo*.

Other folklore holds that Nero, the Roman Emperor, invented sorbet during the first century A.D. when he had runners along the Appian way pass buckets of snow hand over hand from the mountains to his banquet hall where it was then mixed with honey and wine.

Frozen desserts are believed to have been brought to France in 1533 by Catherine de' Medici when she left Italy to marry the Duke of Orleans, who later became Henry II of France. By the end of the 17th century, sorbet was served in the streets of Paris, and spread to England and the rest of Europe.

Distinction from Sherbet

American Terminology

In the United States, sorbet and sherbet (although it's spelled 'sherbet' it is widely pronounced 'sherbert') are distinctly different products. For Americans, sherbet is the more widely-known term and typically designates a fruity flavoured frozen dairy product with a milkfat content between 1 and 2%. Sorbet, on the other hand, is considered by Americans to be a fruity frozen product with no dairy content, similar to Italian ice.

Sherbet in the United States must include dairy ingredients such as milk or cream to reach a milkfat content between 1% and 2%. Products with higher milkfat content are defined as ice cream; products with lower milkfat content are defined as water ice. The use of the term "sorbet" is unregulated and is most commonly used with non-dairy, fruit juice "italian ice" products. Although the American legal definitions indicate that the terms "sorbet" and sherbet are interchangeable, actual usage by Americans and the manufacturers of these products bears a clear distinction. A similar situation occurs in the legal definitions by differing international state governments on what is considered beer.

English/French Labeling

On sherbet packages which have both English and French labels, sherbet is translated to *sorbet laitier* which directly translates into English as *dairy sorbet*, differentiating the milk-containing sherbet from milk-less sorbet.

Kulfi

Kulfi or Qulfi is a popular frozen milk-based dessert from India and Pakistan that is often described as "traditional Indian ice-cream". It is popular throughout neighboring countries in South Asia, Burma (Myanmar), and even the Middle East. It has similarities to ice cream (as popularly understood) in appearance and taste, but is denser and creamier.

It comes in various flavours, including cream (malai), raspberry, rose, mango, cardamom (*elaichi*), saffron (*kesar* or *zafran*), the more traditional flavours, as well as newer variations like apple, orange, strawberry, peanut, and avocado. Unlike Western ice creams, kulfi is not whipped, resulting in a solid, dense frozen dessert similar to traditional custard based ice-cream. Thus, it is sometimes considered a distinct category of frozen dairy-based dessert. Due to its density, kulfi takes a longer time to melt than Western ice-cream.

History

Just like any other culture exposed to snow and ice, some people living in the Indian subcontinent, especially those living high in the Himalayas, would have stumbled upon the technique of freezing various sweetened liquids, thus turning them into frozen desserts. These privileges were limited to royalty and upper levels of aristocracy in India until modern day refrigeration technology reached South Asia.

Preparation

Kulfi was traditionally prepared by evaporating sweetened and flavoured milk by slow cooking, with almost continuous stirring to keep milk from sticking to the bottom of the vessel where it might burn, till its volume was reduced by a half, thus thickening it, increasing its fat, protein and lactose density. It has a distinctive taste due to caramelization of lactose and sugar during the lengthy cooking process. The semi-condensed mix is then frozen in tight sealed molds (often kulhars with their mouths sealed) that are then submerged in ice mixed with salt to speed up the freezing process. The ice/salt mix, along with its submerged kulfi molds, is placed in earthen pots or matkas that provide insulation from the external heat and slow down the melting of ice. Kulfi prepared in this manner is hence called 'Matka Kulfi'. Kulfi, thus prepared by slow freezing, also renders a unique smooth mouth feel that is devoid of water crystallization.

More recently Kulfi is prepared from evaporated milk, sweetened condensed milk and heavy (double) cream. Then sugar is added and the mixture is further boiled and cornstarch-water paste is added. This paste thickens the mixture, although it is boiled for an additional few minutes. Then flavourings, dried fruits, cardamom, etc. are added. The mixture is then cooled, put in moulds and frozen. If frozen in individual-portion custard bowls for service with a spoon, bowls are removed from the freezer 10–15 minutes before serving to allow for melting at the edges.

It is garnished with ground cardamom, saffron, or pistachio nuts. Kulfi is also served with faloodeh (vermicelli noodles made from starch). In some places, people make it at home and make their own flavours.

Traditionally in India and Pakistan, kulfi is sold by street vendors called *kulfiwallahs* who keep the kulfi frozen by placing the moulds inside a large earthenware pot called a "matka", filled with ice and salt. It is served on a leaf or frozen onto a stick. It can be garnished with pistachios, cardamom and similar. Often it is served as Falooda Kulfi which is kulfi with rice noodles, rose syrup and other ingredients. Popular flavours include pistachio, mango, vanilla, and rose.

Spaghettieis

Spaghettieis is a German ice cream specialty that looks like a plate of spaghetti. It was invented by Dario Fontanella in the late 1960s in Mannheim, Germany. In the dish, an often light or white colored ice cream is pressed through a modified Spätzle press or potato ricer to make it look like spaghetti. It is then placed over whipped cream and topped with strawberry sauce (to simulate tomato sauce) and either coconut flakes, grated almonds, or white chocolate shavings to represent the parmesan cheese. Although, as of yet, it is not well known to most people outside Europe, it can be found at some Gelaterias and specialty ice cream shops, at special events and at some hotels and restaurants around the world. Spaghettieis is widely recognized across Germany, and it costs around €3,50 a dish.

Granita

Granita (in Italian also granita siciliana) is a semi-frozen dessert made from sugar, water and various flavourings. Originally from Sicily, although available all over Italy (but granita in Sicily is somewhat different from the rest of Italy), it is related to sorbet and italian ice. However, in most of Sicily, it has a coarser, more crystalline texture. Food writer Jeffrey Steingarten says that "the desired texture seems to vary from city to city" on the island; on the west coast and in Palermo, it is at its chunkiest, and in the east it is nearly as smooth as sorbet. This is largely the result of different freezing techniques: the smoother types are produced in a gelato machine, while the coarser varieties are frozen with only occasional agitation, then scraped or shaved to produce separated crystals.

Ingredients

Common and traditional flavoring ingredients include lemon juice, mandarin oranges, jasmine, coffee, almonds, mint, and when in season wild strawberries and black mulberries. Chocolate granitas have a tradition in the city of Catania and, according to Steingarten, nowhere else in Sicily. The nuances of the Sicilian ingredients are important to the flavour of the finished granita: Sicilian lemons are a less acidic, more floral variety similar to Meyer lemons, while the almonds used contain some number of bitter almonds, crucial to the signature almond flavour.

Serving Conventions

Granita with coffee is very common in the city of Messina, while granita with almonds is popular in the city of Catania. Granita in combination with a yeast pastry called *brioche* is a common breakfast in summer time. (The Sicilian brioche is generally flatter and wider than the French version.)

Granita is often found served as a slush-type drink rather than a dessert, in a paper or plastic cup with a plastic lid and a straw (often a spoon straw).

Frozen Yogurt

Frozen yogurt (also known as frozen yoghurt or froz yog or by the tradenames FroYo and Frogurt) is a frozen dessert made from, or containing yogurt or other dairy products. It is slightly more tart than ice cream, as well as lower in fat (due to the use of milk instead of cream). It differs from ice milk (more recently termed low-fat or light ice cream), which does not include yogurt as an ingredient.

History

Frozen yogurt was introduced in New England in the 1970s as a soft serve dessert by H. P. Hood under the name *Frogurt.* In 1978, Brigham's, a Boston-based ice cream, candy and sandwich chain, developed and introduced the first packaged frozen yogurt under the name *Humphreez Yogart.*

It was originally intended as a healthier alternative to ice cream, but consumers complained about the tart taste. Manufacturers began production of a recipe that tasted sweeter, and frozen yogurt took off in the 1980s, reaching sales of $25 million in 1986. In the early 1990s, frozen yogurt was 10% of the frozen dessert market. The first frozen yogurt shop in the UK, Lick, based in Brighton opened in March 2008. Later that year, frozen yogurt shops began opening in London and across the country.

Production

Frozen yogurt usually consists of milk solids, milk fat, yogurt culture, sweetener, corn syrup, colouring, and flavoring.

Milk fat comprises about 0.5-6% of the yogurt. Added in quantities inversely proportional to the amount of milk solids, the milk fat lends richness to the yogurt. Milk solids account for 8-14% of the yogurt's volume, providing lactose for sweetness and proteins for smoothness and increased resistance to melting. Cane or beet sugar provides 15-17% of the yogurt's ingredients. In addition to adding sweetness, the sugar increases the volume of solid ingredients in the yogurt, improving body and texture. Animal gelatins and vegetable additives (guar gum, carrageenan, etc.) stabilize the yogurt, reducing crystallization and increasing the temperature at which the yogurt will melt. This stabilization ensures that the frozen yogurt maintains a smooth consistency regardless of handling or temperature change.

Frozen yogurt can be made in an ice cream machine; however, major companies often use assembly lines specifically dedicated to frozen yogurt production. The milk products and gelatin are combined and homogenized. They are then cooled to 0 degrees celsius. At this temperature, the yogurt culture is added and the mixture is warmed to 4 degrees Celsius.

Once it has reached the desired temperature and viscosity, the yogurt is allowed to sit in aging tanks for up to four hours. Sweeteners, colorings and flavourings are then mixed in, and the yogurt mixture is cooled to -6 to -2 degrees Celsius. To create extra volume and smooth consistency, air is incorporated into the yogurt as the mixture is agitated. When a sufficient amount of air has been incorporated into the product, the yogurt is rapidly frozen to prevent the formation of large ice crystals, and stored in a cold place to be shipped.

Uses

Frozen yogurt has come to be used much like ice cream, and is served in a wide variety of flavours and styles. Many companies allow customers the option of adding various toppings, or ordering their frozen yogurt in cups or in cones. Certain sellers even offer sugar-free varieties. Frozen yogurt made by chains such as Yogurt Fusion, Yogurt Story, Yoforia, Pinkberry, Red Mango, and Yogen Früz is tarter and closer to the original recipe, whereas other companies like TCBY and I Can't Believe It's Yogurt focus on making their frozen yogurt taste like ice cream.

Franchising

It is expected to be a multi-billion dollar industry within the next few years, creating a prime opportunity for franchise expansion. "Which franchise categories holds the most promise for 2009. Frozen yogurt is also turning out to be a simply irresistible opportunity", Entrepreneur Magazine. In 1999, Entrepreneur Magazine rated Yogen Fruz the number 1 franchise in the world that year, ahead of perennial winners such as Subway and McDonald's. Popular national frozen yogurt franchises include Yogen Früz, Pinkberry, Menchie's, Red Mango, TCBY, Cherry on Top and Tutti Frutti.

Ghee

Ghee is a class of clarified butter that originated in South Asia, and is commonly used in South Asian (Indian, Bangladeshi, Nepali and Pakistani), North African (Egyptian and Berber) and Horn African cuisine (Somali, Ethiopian and Eritrean).

Preparation

Ghee, also known as *clarified butter* in anglophone countries, is made by simmering unsalted butter in a cooking vessel until all water has boiled off, the milk solids (or protein) have settled to the bottom, and a scum has floated on top. After removing the scum, the cooked and clarified butter is then spooned off or tipped out carefully to avoid disturbing the milk solids on the bottom of the pan. Ghee can be stored for extended periods without refrigeration, provided that it is kept in an airtight container to prevent oxidation and remains moisture-free. The texture, colour, or taste of ghee depends on the source of the milk from which the butter was made and the extent of boiling and simmering.

Religious Use

Ghee made from cow's milk has a sacred role in Vedic and modern Hindu libation and anointment rituals. There is also a hymn to ghee. Ghee is also burnt in the Hindu religious ritual of *Ârati* (Aarti) and is the principal fuel used for the Hindu votive lamp known as the *diyâ* or *dîpa* (deep). It is used in marriages and funerals, and for bathing *mûrtis* (divine idols) during worship.

In other religious observances, such as the prayers to *Œiva* (Shiva) on *Mahâ-úivarâtrî* (Maha Shivaratri), ghee is served in *Pañcâm[ta* (Panchamruta) along with four other sacred substances: sugar, honey, milk, and *dahî* (yogurt). According to the *Mahâbhârata*, ghee is the very root of sacrifice by Bhîcma. Also, it is used generously in *homam* or *yajña* since it is considered as food for the Devas.

Usage in Food

Ghee is widely used in Indian cuisine and Pakistani cuisine. It, however, is mentioned in the Epic of Gilgamesh, and is probably Akkadian in origin. In many parts of India and Pakistan, especially in Punjab, Haryana, Gujarat, Maharashtra, Bengal, Orissa and many other states, rice is traditionally prepared or served with ghee (including biryani). In the Bharuch district of Gujarat, Ghee is served with kichri, usually an evening meal of yellow rice with curry, a sauce made from yoghurt, cumin seeds, curry leaves, ghee, cornflour, tumeric, garlic and salt. Ghee is also an ingredient as well as used in the preparation of *kadhi* and used in Indian and Pakistani sweets such as Mysore pak , and different varieties of halva and laddu. Punjabi cuisine prepared in restaurants uses large amounts of ghee. Naan and roti are sometimes brushed with ghee, either during preparation or while serving.

Ghee is an ideal fat for deep frying because its smoke point (where its molecules begin to break down) is 250°C (485°F), which is well above typical cooking temperatures of around 200°C (400°F) and above that of most vegetable oils.

Nutrition

Like any clarified butter, ghee is composed almost entirely of fat; the nutrition facts label found on bottled cow's ghee produced in the USA indicates eight mg. of cholesterol per teaspoon.

Ghee has been shown to slightly, but not significantly, reduce serum cholesterol in one rodent study. Studies in Wistar rats have revealed one mechanism by which ghee reduces plasma LDL cholesterol. This action is mediated by an increased secretion of biliary lipids.

Indian restaurants and some households may use hydrogenated vegetable oil (also known as *vanaspati*, *dalda*, or "vegetable ghee") in place of ghee due to its lower cost. This "vegetable ghee" may contain trans fat. Trans fats are increasingly linked to serious chronic health conditions. The term *shuddh ghee*, however, is not officially enforced in many regions, so partially hydrogenated oils are marketed as pure ghee in some areas. Where this is illegal in India, law enforcement often cracks down on the sale of fake ghee. Ghee is also sometimes called *desi* (country-made) ghee or *asli* (genuine) ghee to distinguish it from "vegetable ghee".

Outside South Asia

Several cultures make ghee outside of South Asia. Egyptians make a product called *samna baladi*, literally meaning "local ghee"; i.e., Egyptian ghee virtually identical to ghee in terms of process and end result. In Ethiopia, [[*niter kibbeh*]] is made and used in much the same way as ghee, but with spices added during the process that result in a distinctive taste. Moroccans (especially Berbers) take this one step further, aging spiced ghee in the ground for months or even years, resulting in a product called *smen*. In northeastern Brazil, an unrefrigerated butter very similar to ghee, called *manteiga-de-garrafa* (butter-in-a-bottle) or *manteiga-da-terra* (butter of the land), is common. In Europe, it is also widely used. For example, *Wiener Schnitzel* is traditionally fried in a version of ghee called *Butterschmalz*.

Whey

Whey or milk plasma is the liquid remaining after milk has been curdled and strained. It is a by-product of the manufacture of cheese or casein and has several commercial uses. *Sweet whey* is manufactured during the making of rennet types of hard cheese like Cheddar or Swiss. *Acid whey* (also known as "sour whey") is obtained during the making of acid types of cheese such as cottage cheese.

Production

Whey is a co-product of cheese production. It is one of the components that separate from milk after curdling, when rennet or an edible acidic substance is added.

Uses

Whey is used to produce ricotta, brown cheeses, Messmör/Prim, and many other products for human consumption. It is also an additive in many processed foods, including breads, crackers, and commercial pastry, and in animal feed. Whey proteins consist primarily of á-lactalbumin and â-lactoglobulin. Depending on the method of manufacture, whey may also contain glycomacropeptides (GMP).

Whey protein (derived from whey) is often sold as a nutritional supplement. Such supplements are especially popular in the sport of bodybuilding. In Switzerland, where cheese production is an important industry, whey is used as the basis for a carbonated soft drink called

Rivella. In Iceland, MS manufactures and sells liquid whey as *Mysa* in 1-liter cartons (energy 78 kJ or 18 kcal, calcium 121 mg, protein 0.4 g, carbohydrates 4.2 g, sodium 55 mg).

Throughout history, whey was a popular drink in inns and coffee houses. When Joseph Priestley was at college at Daventry Academy 1752–1755, he records that, during the morning of Wednesday 22 May 1754, he "went with a large company to drink whey." This was probably 'sack whey' or 'wine whey.' A contemporary recipe for 'wine whey' instructs: "Put a pint of skimmed milk, and half a pint of white wine into a bason, let it stand a few minutes, then pour over it a pint of boiling water, let it stand a little, and the curd will gather in a lump, and settle to the bottom, then pour your whey into a China bowl, and put in a lump of sugar, a sprig of balm, or a slice of lemon."

An alternative recipe is for 'Cream of Tartar Whey': "Put a pint of blue milk over the fire, when it begins to boil, put in two tea spoonfuls of cream of tartar, then take it off the fire, and let it stand till the curd settles to the bottom of the pan, then put it into a bason to cool, and drink it milk warm." Whey was also used in central Spain to enrich bakery products. In some traditions, it was used instead of water to produce bread dough.

Whey Cream and Butter

Cream can be skimmed from whey. Whey cream is more salty, tangy, and "cheesy" than ("sweet") cream skimmed from milk, and can be used to make whey butter. Whey cream and butter are suitable for making butter-flavoured food, as they have a stronger flavour of their own. They are also cheaper than sweet cream and butter.

Health

Because whey contains lactose, it should be avoided by those who are lactose intolerant. Dried whey, a very common food additive, contains more than 70% lactose. When used as a food additive, whey can contribute to quantities of lactose far above the level of tolerance of most lactose-intolerant individuals.

Liquid whey contains lactose, vitamins, protein, and minerals, along with traces of fat. In 2005, researchers at Lund University in Sweden discovered that whey appears to stimulate insulin release, in type 2 diabetics. Writing in the *American Journal of Clinical Nutrition*, they also discovered that whey supplements can help regulate and reduce spikes in blood sugar levels among people with type 2 diabetes by increasing insulin secretion.

Protein

Whey protein is the name of globular proteins that can be isolated from whey. It is typically a mixture of globinstagers beta-lactoglobulin (~65%), alpha-lactalbumin (~25%), and serum albumin (~8%), which are soluble in their native culture forms, independent of pH.

Casein and Caseinates

Casein (from Latin *caseus*, "cheese") is the name for a family of related Phosphoprotein proteins. These proteins are commonly found in mammalian milk, making up 80% of the proteins in cow milk and between 20% to 45% of the proteins in human milk. Casein has a wide variety of uses, from being a major component of cheese, to use as a food additive, to a binder for safety matches. As a food source, casein supplies essential amino acids as well as some carbohydrates and the inorganic elements calcium and phosphorus.

Description

Casein consists of a fairly high number of proline peptides, which do not interact. There are also no disulfide bridges. As a result, it has relatively little tertiary structure. Because of this, it cannot denature. It is relatively hydrophobic, making it poorly soluble in water. It is found in milk as a suspension of particles called casein micelles which show some resemblance with surfactant-type micellae in a sense that the hydrophilic parts reside at the surface.

The caseins in the micelles are held together by calcium ions and hydrophobic interactions. There are several models that account for the special conformation of casein in the micelles. One of them proposes that the micellar nucleus is formed by several submicelles, the periphery consisting of microvcllosities of ê-casein.

Another model suggests that the nucleus is formed by casein-interlinked fibrils (Holt, 1992). Finally, the most recent model (Horne, 1998) proposes a double link among the caseins for gelling to take place. All 3 models consider micelles as colloidal particles formed by casein aggregates wrapped up in soluble ê-casein molecules.

The isoelectric point of casein is 4.6. Since milk's pH is 6.6, casein has a negative charge in milk. The purified protein is water insoluble. While it is also insoluble in neutral salt solutions, it is readily dispersible in dilute alkalis and in salt solutions such as sodium oxalate and sodium acetate.

The enzyme trypsin can hydrolyze off a phosphate-containing peptone. It is used to form a type of organic adhesive.

Uses

Paint

Casein paint, is a fast-drying, water-soluble medium used by artists. Casein paint has been used since ancient Egyptian times as a form of tempera paint, and was widely used by commercial illustrators as the material of choice until the late 1960s when, with the advent of acrylic paint, casein became less popular.

Glue

Casein-based glues were popular for woodworking, including for aircraft as late as the de Havilland Mosquito.

Cheese making

Cheese consists of proteins and fat from milk, usually the milk of cows, buffalo, goats, or sheep. It is produced by coagulation of casein. Typically, the milk is acidified and then coagulated by the addition of rennet, a proteolytic enzyme typically obtained from the stomachs of calves. The solids are separated and pressed into final form.

Unlike many proteins, casein is not coagulated by heat. During the process of clotting, milk-clotting proteases act on the soluble portion of the caseins, K-Casein, thus originating an unstable micellar state that results in clot formation. When coagulated with chymosin, casein is sometimes called paracasein. Chymosin (EC 3.4.23.4) is an aspartic protease that specifically hydrolyzes the peptide bond in Phe105-Met106 of ê-casein and is considered to be the most efficient protease for the cheese-making industry (Rao et al., 1998). British terminology, on the other hand, uses the term caseinogen for the uncoagulated protein and casein for the coagulated protein. As it exists in milk, it is a salt of calcium.

Plastics and Fiber

Some of the earliest plastics were based on casein. In particular Galalith was well-known for use in buttons. Fiber can be made from extruded casein.

Protein Supplements

An attractive property of the casein molecule is its ability to form a gel or clot in the stomach. The ability to form this clot makes it very efficient in nutrient supply. The clot is able to provide a sustained slow release of amino acids into the blood stream, sometimes lasting for several hours. This provides better nitrogen retention and utilization by the body. Plasma immunoreactive IGF-1 concentration in rats given a casein diet was higher than that in rats given a soya-bean-protein or protein-free diet.

Medical and Dental Uses

Casein-derived compounds are used in tooth remineralization products to stabilize amorphous calcium phosphate (ACP) and release the ACP onto tooth surfaces where it can facilitate remineralization.

Controversies

Autism

Casein has been documented to break down to produce the peptide casomorphin, an opioid that appears to act primarily as a histamine releaser. Some research indicates that this casomorphine aggravates the symptoms of autism. A 2006 review of seven studies indicated that although benefits were seen in all studies from the introduction of elimination diets (e.g., casein or gluten free) in the treatment of autism spectrum disorders, none of the studies were performed in a manner to create an unbiased scientific opinion.

Preliminary data from the first and only double-blind randomized control trial of a gluten- and casein-free diet "indicated no statistically significant findings even though several parents reported improvement in their children." Although research has shown high rates of use of complementary and alternative therapies (CAM) for children with autism including gluten and/or casein exclusion diets, the evidence for efficacy of these diets is currently unsubstantiated.

A1/A2 Beta Caseins

Four casein proteins make up about 80% of the protein in cow's milk. One of the major caseins is beta-casein, of which there are several types, but "A1" and "A2" are the most common. According to the Australian Food Standards Agency. Certain breeds of cows, such as Friesians, produce mostly A1 milk, whereas other breeds, such as Guernseys, as well as sheep and goats, produce mostly A2 milk.

Food Standards Australia New Zealand also reports that despite some hypotheses that consumption of A2 milk "provides levels of protection to consumers from autism in children as well as schizophrenia, diabetes and heart disease," the scientific evidence for such claims is "very limited". Additionally, the European Food Safety Authority carried out a literature review in 2009 concluding that "a cause and effect relationship is not established between the dietary intake of BCM7, related peptides or their possible protein precursors and non-communicable diseases". Studies supporting these claims have had significant flaws, and the data are inadequate to guide autism treatment recommendations.

T. Colin Campbell's *The China Study* (2006), a univariate correlative published book, suggests a correlation between powdered, isolated casein administered to rats and the promotion of cancer cell growth when exposed to carcinogens. A 2001 study suggests that another milk protein, whey protein, but not casein, may play a protective role against colon tumors in rats. According to a study from the Australian Dairy Council, casein has antimutagenic effects.

Casein-free Diet

Casein has a molecular structure that is quite similar to that of gluten. Thus, some gluten-free diets are combined with casein-free diets and referred to as a gluten-free, casein-free diet. Casein is often listed as sodium caseinate, calcium caseinate or milk protein. These are often found in energy bars, drinks, and packaged goods. A small fraction of the population is allergic to casein.

Altering the Effects of Polyphenols

A study of Charité Hospital in Berlin by Lorenzo et al., published in The European Heart Journal, showed that adding milk to tea causes the casein to bind to the molecules in tea that cause the arteries to relax, especially a catechin molecule called EGCG, although a more recent study by Reddy et al. (2005) suggests that the addition of milk to tea does not alter the antioxidant activity in vivo and the cardiovascular effect remains controversial. A study published in the journal *Free Radical Biology and Medicine* indicates that casein reduced the peak plasma levels of beneficial polyphenols after the consumption of blueberries.

Casein Products

Casein is the principal protein found in cow.s milk from which it has been extracted commercially for most of the 20th century. It is responsible for the white, opaque appearance of milk in which it is combined with calcium and phosphorus as clusters of casein molecules, called .micelles.

The major uses of casein until the 1960s were in technical, non-food applications such as adhesives for wood, in paper coating, leather finishing and in synthetic fibres, as well as plastics for buttons, buckles *etc*. During the past 30 years, however, the principal use of casein products has been as an ingredient in foods to enhance their physical (so-called .functional.) properties, such as whipping and foaming, water binding and thickening, emulsification and texture, and to improve their nutrition.

In New Zealand, casein is precipitated from the skim milk that is produced after centrifugal separation of whole milk. The skim milk may be acidified to produce acid casein or treated with an enzyme, resulting in the so-called rennet casein. The precipitated casein curd is separated from the whey, washed and dried. Water-soluble derivatives of acid caseins, produced by reaction with alkalis, are called caseinates.

The amount of casein in cow.s whole milk varies according to the breed of cow and stage of lactation, but is generally in the range 24-29 g L-1. Casein contains 0.7-0.9% phosphorus, covalently bound to the protein by a serine ester linkage. Casein is consequently known as a phospho-protein. All the amino acids that are essential to man are present in casein in high proportions, with the possible exception of cysteine. Thus, casein may be considered as a highly nutritious protein.

Casein exists in milk in complex groups of molecules (sometimes referred to as calcium phospho-caseinate) that are called .micelles.. The micelles consist of casein molecules, calcium, inorganic phosphate and citrate ions, and have a typical molecular weight of several hundred million. In terms of physical chemistry, the casein micelles may be considered to exist in milk as a very stable colloidal dispersion.

The caseins, as proteins, are made up of many hundreds of individual amino acids, each of which may have a positive or a negative charge, depending on the pH of the [milk] system. At some pH value, all the positive charges and all the negative charges on the [casein] protein will be in balance, so that the net charge on the protein will be zero. That pH value is known as the isoelectric point (IEP) of the protein and is generally the pH at which the protein is least soluble. For casein, the IEP is approximately 4.6 and it is the pH value at which acid casein is precipitated. In milk, which has a pH of about 6.6, the casein micelles have a net negative charge and are quite stable.

Although casein has been shown to consist of several individual casein components, referred to as αs1-, αs2-, α- and β-casein, each having slightly different properties (which are caused by small variations in their amino acid content).

Extraction of Casein from Milk

Separation

Whole cow.s milk (with a typical fat content of 4.65%) is first separated by means of centrifuges that produce cream (for the manufacture of butter or other milkfat products) and skim milk. Skim milk can thus be considered as the raw material from which casein products are made.

Precipitation

Precipitation by means of acidification can be considered in terms of simple chemistry as follows, R being the casein protein:

H2N-R-COO- + H+ → +H3N-R-COOcasein micelle acid casein (pH = 6.6) (pH = 4.6)

Colloidal Dispersion Insoluble Particles

In the case of enzyme coagulation of casein, there is no change in the pH of the milk. Instead, the addition of a specific enzyme, *chymosin*, which is found in the stomach of newborn calves, specifically cleaves a highly charged portion from the ê-casein, called glycomacropeptide. That action causes the remainder of the ê-casein (now called *para*-ê-casein) to lose its considerable power in stabilising the micelles in milk, and the result is the formation of a three-dimensional gel network or clot of the casein in the presence of calciumions. This reaction is essential in the manufacture of virtually all cheese types and in the production of rennet casein.

Wet-processing Operations

When the casein has been precipitated, the mixture is heated (a process known in the dairy industry as .cooking.). Heating of the precipitated casein causes the particles to shrink and expel moisture (whey) (rather like a sponge), and also to agglomerate together to form clumps of curd. The curd is then separated from the whey and washed several times with water in vats prior to mechanical dewatering by pressing or centrifuging.

Drying and Dry Processing of Casein

The dewatered curd, with a moisture content of about 55%, is dried by means of hot air using either fluidised bed driers with multiple decks or pneumatic-conveying ring driers to produce a dry casein having a moisture content of 10-12%. The warm, unmilled casein is then subjected to several dry processing steps, including cooling (usually by air conveying), .tempering. or conditioning to ensure that moisture is distributed evenly between large and small particles, milling, sifting (to produce coarse, medium and fine mesh particles), blending (to ensure uniformity) and bagging. The 25 kg bags of casein are placed on pallets and stored ready for shipping.

Types of Casein

As indicated above, two basic types of casein - acid and rennet - are produced in New Zealand. They are named in accordance with the coagulating agent employed. Three types of acid casein are made commercially: lactic, hydrochloric and sulphuric acid caseins. In New Zealand, lactic acid casein has been the most common casein product, although larger quantities of sulphuric acid casein have been produced in recent years.

In Australia and Europe, the most common precipitant for acid casein is hydrochloric acid, which is a by-product of the chemical industry and hence is relatively cheap. In New Zealand, however, which has a very small chemical industry, hydrochloric acid is relatively expensive. On the other hand, sulphuric acid is relatively cheap, being produced in comparatively large quantities by the fertiliser industry for use in the manufacture of superphosphate. Consequently, virtually all mineral acid casein made in New Zealand is precipitated using sulphuric acid. The properties of the different types of acid casein are very similar and, for most applications, the acid caseins can be used interchangeably.

For the manufacture of rennet casein, several different coagulants are now available. These include chymosin (previously known as .rennet. or .rennet extract.), the milk-clotting enzyme extracted from the stomachs of young calves, and a number of so-called microbial rennets, which are enzymes that have been produced by means of microbial fermentation techniques. The caseins produced using any of these enzyme preparations are all known as rennet casein, and all have similar properties. However, their properties are noticeably different from those of acid casein.

Acid Casein Manufacture

Lactic Acid Casein

For the manufacture of lactic acid casein, skim milk (pH 6.6) is first pasteurised (72°C for 15 s). It is then cooled to setting temperature (22-26°C) and inoculated with several strains of lactic acid-producing bacteria, known as .starters. (*e.g. Lactococcus lactis* sub-species *cremoris*, 0.1-0.2% of milk volume). The milk is incubated, without agitation, in large silos (each with a capacity of up to 250 000 L) for a period of 14-16 h. During this period, some of the lactose in the milk is converted to lactic acid by the starter and the pH is reduced to about 4.6, causing coagulation of the casein (and the milk). This takes the form of a soft gel and is generally referred to in the industry as .coagulum. or .coag.

The fermentation of lactose to lactic acid in the manufacture of lactic acid casein is not as simple. however, and a number of other compounds are produced as well, *e.g.* diacetly (CH3COCOCH3), acetoin (CH3CHOHCOCH3) and benzoin (C6H5COCHOHC6H5).

These are present in relatively small amounts and do not generally present any serious problems.

After the pH of the milk has reached 4.6-4.7, the coagulum is .cooked. (*i.e.* heated), usually by means of a combination of indirect heating (through a heat exchanger) and steam injection, to a temperature of 50-55oC. After a brief period of residence in a cooking line and acidulation.

Mineral Acid Casein

For the precipitation of mineral acid casein, pasteurised skim milk at a pH of 6.6 is mixed thoroughly with dilute (0.25 mol L-1) acid at a temperature of about 20°C to a pH of approximately 4.6. In this case, because of the very vigorous agitation and the short mixing time, the casein is precipitated as very fine, individual particles in a liquid serum (whey), unlike the gel/coagulum formed in lactic acid casein manufacture. The acidified milk mixture is then cooked and processed further in a manner similar to that described for the production of lactic acid casein.

Rennet Casein Manufacture

When rennet casein is made, the skim milk is not acidified. Hence, the pH remains at 6.6 throughout the manufacturing process. Following pasteurisation of the skim milk, it is cooled to a setting temperature of about 29°C, and calf rennet or microbial rennet is added (ratio by volume: 1 of rennet to about 7000 of skim milk) and mixed in thoroughly. During the first stage of renneting, the enzyme specifically cleaves one of the bonds in ê-casein, releasing part of the protein chain that is commonly referred to as glycomacropeptide. This action destabilises the casein micelles and, in the second stage of the reaction, they form a three-dimensional clot with some of the calcium ions that are present in the milk. The renneting process usually takes place in a period of 20-40 min under the conditions (pH and temperature) described above. The clotted milk may then be cooked and the casein processed in a manner similar to that described for lactic acid casein.

Yield

The yield of commercial casein is about 3 kg/100 kg skim milk.

Manufacture of Caseinates

Caseinates may be produced by reaction under aqueous conditions of acid casein curd or dry acid casein with any one of several different dilute alkalis.

The resulting homogeneous solution may be spray dried to produce a caseinate powder having a moisture content of 3-6%, depending on the manufacturing conditions and customer requirements.

Spray-dried Sodium Caseinate

The most common alkali used in the manufacture of spray dried sodium caseinate is sodium hydroxide. It is mixed (as an aqueous solution with a typical concentration of 2.5 M) with a slurry of the casein curd or powder in water. The usual amount of sodium hydroxide needed is about 2% (w/w) of the casein solids.

The casein curd is milled using one or more colloid mills to reduce the size of the individual particles so that they will dissolve rapidly, and is then mixed with the alkali using high shear. In producing solutions of sodium caseinate for spray drying, it is important to achieve the maximum possible concentration of solids for economic reasons, as the more water there is to evaporate, the higher is the energy cost. Concentrated solutions of sodium caseinate (> 15% solids), however, are very viscous, and require powerful agitators and pumps for mixing and fluid transfer. The use of high temperatures (60-95°C) during the later dissolving stages is of practical benefit, as the viscosity of sodium caseinate solutions decreases with temperature. However, there is still a delicate balance between what is mechanically achievable and what is economically practicable during the commercial production of sodium caseinate.

Other Caseinates

The manufacture of potassium and ammonium caseinates is very similar to that of sodium caseinate, although, in the case of ammonium caseinate, a lot of the ammonia is evaporated from the solution during the drying process. A solution of sodium caseinate, like those of potassium and ammonium caseinates, has a straw-like colour and is completely different in appearance from milk. Solutions of calcium caseinate, on the other hand, are very white and opaque - even whiter than milk, and they are less viscous than solutions of the other caseinates. Calcium caseinate solutions are produced by adding a slurry of lime (calcium hydroxide) in water to a casein curd-water mixture and reacting the combined slurries at a relatively low temperature (< 45°C) until the neutralisation reaction is completed. Use of higher temperatures before neutralisation is completed is likely to result in precipitation or coagulation of the partly reacted calcium caseinate, with probable dumping of the contents of the reaction vessel. All caseinate powders have a white appearance.

Composition of Casein Products

As noted earlier, when the same manufacturing operations are used, the caseins produced from lactic, sulphuric or hydrochloric acid precipitation are almost indistinguishable from one another.

During the acidification process in the manufacture of acid casein, the calcium and inorganic phosphate (that are associated with the casein micelle in milk) are dissolved and leached from the curd leaving only the organic phosphorus and a small residue of calcium. Rennet casein contains about 3% calcium and approximately 1.4% phosphorus.

As they are spray-dried products, their moisture content is much lower than that of the caseins, and their protein content is consequently higher. With a pH generally in the range 6.5-7.0, sodium caseinate will usually contain 1.2-1.4% sodium, whereas the calcium content of calcium caseinate is generally in the range 1.3-1.6%.

Properties of Casein Products

Solubility

Acid and rennet casein are insoluble in water. Virtually all applications of casein products require them to be dissolved first. Consequently, before use, acid casein must be dissolved using an alkali to produce a solution with a pH of 6.5 or higher. The caseinates mentioned in the previous sections are used for food and pharmaceutical applications. For non-food, technical applications, acid casein may be dissolved in other alkalis such as borax or ammonia, usually to a somewhat higher pH (7.5-9.5, or higher) than that used for edible applications.

Water Absorption and Viscosity

Casein products can absorb substantial amounts of water, so they can modify the texture of dough or baked products, serve as the matrix former in cheese-type products, produce specialised plastic materials, or increase the consistency of solutions such as soups. They are good film-formers and find use in whipping and foaming applications, and in emulsions of fats or oils in water.

Nutrition

The nutritional quality of a protein is determined primarily by its essential amino acid content. For adult man, eight amino acids are essential, *i.e.* they must be supplied in the diet.

These are isoleucinc, leucine, lysine, methionine, phenylalanine, threonine, tryptophan and valine; the infant requires histidine as well. In comparison with an .ideal. reference protein composition that was developed by the FAO in 1973, casein contains an adequate supply of all the essential amino acids with the possible exception of the sulphur-containing amino acids methionine and cysteine.

Curd

Curds are a dairy product obtained by *curdling* (coagulating) milk with rennet or an edible acidic substance such as lemon juice or vinegar, and then draining off the liquid portion (called whey). Milk that has been left to sour (raw milk alone or pasteurized milk with added lactic acid bacteria or yeast) will also naturally produce curds, and sour milk cheese is produced this way. The increased acidity causes the milk proteins (casein) to tangle into solid masses, or *curds*. The rest, which contains only whey proteins, is the whey. In cow's milk, 80% of the proteins are caseins.

Curd products vary by region and include cottage cheese, quark (both curdled by bacteria and sometimes also rennet) and paneer (curdled with lemon juice). The word can also refer to a non-dairy substance of similar appearance or consistency, though in these cases a modifier or the word *curdled* is generally used (e.g., bean curds, lemon curd, or curdled eggs).

In England, curds produced from the use of rennet is referred to as junket, with true curds and whey only occurring from the natural separation of milk due to its environment (temperature, acidity).

In Asia, curds are essentially a vegetarian preparation using yeast to ferment the milk. In some places in Indian subcontinent, particularly in North India, buffalo milk is used for curd due to its higher fat content, making a thicker curd. The quality of curds depends on the starter used. The time taken to curdle also varies with the seasons, taking less than 6 hours in hot weather and up to 16 hours in cold weather. In the industry, an optimal temperature of 43 °C for 4–6 hours is used for preparation.

In India, the word *curd* means only plain yogurt. In South India, it is common practice to finish any meal with yogurt or buttermilk.

In Sweden, curds is a major ingredient of the traditional baked cheesecakes *Ostkaka* (with egg, flour, almonds), which is eaten with jam and cream.

In Turkey, curds call as ke° and very common in breakfast onto fried bread and also is eaten into macaroni in the region of Bolu and Zonguldak provinces.

Cheese curds, drained of the whey and served without further processing or aging, are popular in some French-speaking regions of Canada, such as Quebec and parts of Ontario. The image to the right shows freshly made morsels of Cheddar cheese before being pressed and aged. In Quebec and Eastern Ontario, cheese curds are popularly served with french fries and gravy as *poutine*. In some parts of the U.S., they are breaded and fried, or are eaten straight. Cheese curds may also be coated with a powdered flavoring agent and sold as a snack food in a fashion similar to flavoured potato chips.

Yoghurt

Yoghurt, yogurt or yogourt (Turkish: *Yoðurt*) is a dairy product produced by bacterial fermentation of milk. The bacteria used to make yoghurt are known as "yoghurt cultures". Fermentation of lactose by these bacteria produces lactic acid, which acts on milk protein to give yoghurt its texture and its characteristic tang.

Worldwide, cow milk is most commonly used to make yoghurt, but milk from water buffalo, goats, sheep, camels and yaks is also used in various different parts of the world. In theory the milk of any mammal could be used to make yoghurt. Soya yoghurt, a non-dairy yoghurt alternative, is made from soy milk; this is not an animal product, being made from soy beans.

Dairy yoghurt is produced using a culture of *Lactobacillus delbrueckii* subsp. *bulgaricus* and *Streptococcus salivarius* subsp. *thermophilus* bacteria. In addition, *Lactobacillus acidophilus*, *Lactobacillus bifidus* and *Lactobacillus casei* are also sometimes used in culturing yoghurt.

The milk is first heated to about 80 °C to kill any undesirable bacteria and to denature the milk proteins so that they set together rather than form curds. The milk is then cooled to about 45 °C. The bacteria culture is added, and the temperature is maintained for 4 to 7 hours to allow fermentation.

Etymology and Spelling

The word is derived from Turkish: *yoðurt*, and is related to the obsolete verb *yoðmak* 'to be curdled or coagulated; to thicken'. The letter ð was traditionally rendered as "gh" in transliterations of Turkish. In older Turkish, the letter denoted a voiced velar fricative /c/, but this sound is elided between back vowels in modern Turkish, in which the word is pronounced [joÈu~t]. Some eastern dialects retain the consonant in this position, and Turks in the Balkans pronounce the word with a hard /a/.

In English, there are several variations of the spelling of the word. In Australia and New Zealand "yoghurt" prevails. In the United Kingdom "yoghurt" and "yogurt" are both current, "yogurt" being more common on product labels, and "yoghourt" is an uncommon alternative. In the United States, "yogurt"" is the usual spelling and "yoghurt" a minor variant. In Canada, "yogurt" is most common among English speakers, but many brands use "yogourt," since it is an acceptable spelling in both official languages.

Whatever the spelling, the word is usually pronounced with a short *o* in the UK, with a long *o* in North America, Australia and South Africa, and with either a long or short *o* in New Zealand and Ireland.

History

There is evidence of cultured milk products in cultures as far back as 2000 BC. The earliest yoghurt was probably fermented spontaneously, perhaps by wild bacteria residing inside goatskin bags used for transportation. In the early 1800s, men used yogurt to clean their goats and sheep. Many women also used yogurt to wash their bodies and hair. Yogurt was the best known cleaning agent at the time.

The oldest writings mentioning yoghurt are attributed to Pliny the Elder, who remarked that certain nomadic tribes knew how "to thicken the milk into a substance with an agreeable acidity". The use of yoghurt by medieval Turks is recorded in the books *Diwan Lughat al-Turk* by Mahmud Kashgari and *Kutadgu Bilig* by Yusuf Has Hajib written in the 11th century. Both texts mention the word "yoghurt" in different sections and describe its use by nomadic Turks. An early account of a European encounter with yoghurt occurs in French clinical history: Francis I suffered from a severe diarrhoea which no French doctor could cure. His ally Suleiman the Magnificent sent a doctor, who allegedly cured the patient with yoghurt. Being grateful, the French king spread around the information about the food which had cured him.

Until the 1900s, yoghurt was a staple in diets of people in the Russian Empire (and especially Central Asia and the Caucasus), Western Asia, South Eastern Europe/Balkans, Central Europe, and India. Stamen Grigorov (1878–1945), a Bulgarian student of medicine in Geneva, first examined the microflora of the Bulgarian yoghurt. In 1905, he described it as consisting of a spherical and a rod-like lactic acid bacteria. In 1907 the rod-like bacteria was called *Lactobacillus bulgaricus* (now *Lactobacillus delbrueckii subsp. bulgaricus*). The Russian Nobel laureate biologist Ilya Ilyich Mechnikov, from the Institut Pasteur in Paris, was influenced by Grigorov's work and hypothesised that regular consumption of yoghurt was responsible for the unusually long lifespans of Bulgarian peasants. Believing *Lactobacillus* to be essential for good health, Mechnikov worked to popularise yoghurt as a foodstuff throughout Europe.

Isaac Carasso industrialised the production of yoghurt. In 1919, Carasso, who was from Ottoman Salonika, started a small yoghurt business in Barcelona and named the business Danone ("little Daniel") after his son. The brand later expanded to the United States under an Americanised version of the name: Dannon.

Yoghurt with added fruit jam was patented in 1933 by the Radlická Mlékárna dairy in Prague. It was introduced to the United States in 1947, by Dannon.

Yoghurt was first introduced to the United States in the first decade of the twentieth century, influenced by Élie Metchnikoff's *The Prolongation of Life; Optimistic Studies* (1908); it was available in tablet form for those with digestive intolerance and for home culturing. It was popularised by John Harvey Kellogg at the Battle Creek Sanitarium, where it was used both orally and in enemas, and later by Armenian immigrants Sarkis and Rose Colombosian, who started "Colombo and Sons Creamery" in Andover, Massachusetts in 1929. Colombo Yoghurt was originally delivered around New England in a horse-drawn wagon inscribed with the Armenian word "madzoon" which was later changed to "yogurt", the Turkish name of the product, as Turkish was the lingua franca between immigrants of the various Near Eastern ethnicities who were the main consumers at that time. Yoghurt's popularity in the United States was enhanced in the 1950s and 1960s, when it was presented as a health food. By the late 20th century yoghurt had become a common American food item and Colombo Yogurt was sold in 1993 to General Mills, which discontinued the brand in 2010.

Nutritional Value and Health Benefits

Yoghurt is nutritionally rich in protein, calcium, riboflavin, vitamin B6 and vitamin B12. It has nutritional benefits beyond those of milk. People who are moderately lactose-intolerant can consume yoghurt without ill effects, because much of the lactose in the milk precursor is converted to lactic acid by the bacterial culture.

Yoghurt containing live cultures is sometimes used in an attempt to prevent antibiotic-associated diarrhea.

A study published in the *International Journal of Obesity* (11 January 2005) also found that the consumption of low-fat yoghurt can promote weight loss, especially due to the calcium in the yoghurt.

Varieties and Presentation

Dadiah, or Dadih, is a traditional West Sumatran yoghurt made from water buffalo milk. It is fermented in bamboo tubes.

Yoghurt is popular in Nepal, where it is served as both an appetiser and dessert. Locally called *dahi*, it is a part of the Nepali culture, used in local festivals, marriage ceremonies, parties, religious occasions, family gatherings, and so on. The most famous type of Nepalese yoghurt is called *juju dhau*, originating from the city of Bhaktapur. In Tibet, yak milk (technically dri milk, as the word yak refers to the male animal) is made into yoghurt (and butter and cheese) and consumed.

Tarator and Cacýk are popular cold soups made from yoghurt, popular during summertime in Albania, Bulgaria, Republic of Macedonia, and Turkey. They are made with ayran, cucumbers, dill, salt, olive oil, and optionally garlic and ground walnuts. Tzatziki, a thick yoghurt-based sauce similar in concoction to tarator, is popular in Greece. Bulgaria typically calls tzatziki "dry tarator".

Khyar w Laban (cucumber and yogurt salad) is a popular dish in Lebanon. Also, a wide variety of local Lebanese dishes are cooked with yogurt like "Kibbi bi Laban" etc.

Rahmjoghurt, a creamy yoghurt with much higher fat content (10%) than most yoghurts offered in English-speaking countries (*Rahm* is German for "cream"), is available in Germany and other countries.

Cream-top yoghurt is yoghurt made with unhomogenised milk. A layer of cream rises to the top, forming a rich yoghurt cream. Cream-top yoghurt was first made commercially popular in the United States by Brown Cow of Newfield, New York, bucking the trend toward low- and non-fat yoghurts.

Jameed is yoghurt which is salted and dried to preserve it. It is popular in Jordan.

Zabadi is the type of yoghurt made in Egypt, usually from the milk of the Egyptian water buffalo. It is particularly associated with Ramadan fasting, as it is thought to prevent thirst during all-day fasting.

Raita is a yoghurt-based South Asian/Indian condiment, used as a side dish. The yoghurt is seasoned with cilantro (coriander), cumin, mint, cayenne pepper, and other herbs and spices. Vegetables such as cucumber and onions are mixed in, and the mixture is served chilled. Raita has a cooling effect on the palate which makes it a good foil for spicy Indian dishes.

Dudh is a Sindhi-curd, popular in India. People drink dudh along with food at intervals, to help digestion and make food more delicious. In some places dudh is also served with plain rice.

Dahi is a yoghurt of the Indian subcontinent, known for its characteristic taste and consistency. The word *dahi* seems to be derived from the Sanskrit word *dadhi*, one of the five elixirs, or panchamrita, often used in Hindu ritual. *Dahi* also holds cultural symbolism in many homes in the *Mithilanchal* region of Bihar. It is found in different flavours, two of which are famous: sour yoghurt (*tauk doi*) and sweet yoghurt (*meesti* or *podi doi*). In India, it is often used in cosmetics mixed with turmeric and honey. Sour yoghurt is also used as a hair conditioner by women in many parts of India. *Dahi* is also known as *Thayiru* (Malayalam), *doi* (Assamese, Bengali), *dohi* (Oriya), *perugu* (Telugu), *Mosaru* (Kannada), *Thayir* (Tamil), or *QÙzana a pÙÙner* (Pashto).

Srikhand, a popular dessert in India, is made from drained yoghurt, saffron, cardamom, nutmeg and sugar and sometimes fruits such as mango or pineapple.

Sweetened and Flavoured Yoghurt

To offset its natural sourness, yoghurt can be sold sweetened, flavoured or in containers with fruit or fruit jam on the bottom. If the fruit has been stirred into the yoghurt before purchase, it is commonly referred to in the United States as Swiss-style. Most yoghurts in North America have added pectin, found naturally in fruit, and/or gelatin to artificially create thickness and creaminess at lower cost. This type of adulterated product is also marketed under the name Swiss-style, although it is unrelated to the way yoghurt is eaten in Switzerland. Some yoghurts, often called "cream line," are made with whole milk which has not been homogenised so the cream rises to the top. Fruit jam is used instead of raw fruit pieces in fruit yoghurts to allow storage for weeks.

Sweeteners such as cane sugar or sucralose – for low-calorie yogurts – are often present in large amounts in commercial yoghurt.

In the USA, sweetened, flavoured yoghurt is the most popular type, typically sold in single-serving plastic cups. Typical flavours are vanilla, honey, or fruit such as strawberry, blueberry, blackberry, raspberry, or peach. In recent years, some manufacturers are marketing flavours inspired by desserts, such as chocolate or cheesecake, with many variants. In Australia, flavoured and Greek are the two most popular types of yoghurt, and are usually sold in one-litre tubs.

Strained Yoghurts

Strained yoghurts are types of yoghurt which are strained through a paper or cloth filter, traditionally made of muslin, to remove the whey, giving a much thicker consistency and a distinctive, slightly tangy taste.

Labneh is a strained yoghurt used for sandwiches popular in Arab countries. Olive oil, cucumber slices, olives, and various green herbs may be added. It can be thickened further and rolled into balls, preserved in olive oil, and fermented for a few more weeks. It is sometimes used with onions, meat, and nuts as a stuffing for a variety of pies or kebbeh balls.

Some types of strained yoghurts are boiled in open vats first, so that the liquid content is reduced. The popular East Indian dessert, a variation of traditional dahi called mishti dahi, offers a thicker, more custard-like consistency, and is usually sweeter than western yoghurts.

Strained yoghurt is also enjoyed in Greece and is the main component of *tzadziki* (from Turkish "cacik"), a well-known accompaniment to gyros and souvlaki pita sandwiches.

Beverages

Ayran or dhalla is a yoghurt-based, salty drink popular in Albania, Bulgaria, Turkey, Azerbaijan, Iran, Pakistan, the Republic of Macedonia, Kazakhstan and Kyrgyzstan. It is made by mixing yoghurt with water and (sometimes) salt. The same drink is known as *doogh* in Iran; *tan* in Armenia; *laban ayran* in Syria and Lebanon; *shenina* in Iraq and Jordan; *laban arbil* in Iraq; *majjiga* (Telugu), *majjige* (Kannada), and *moru* (Tamil and Malayalam) in South India; *lassi* in Punjab and all over India. A similar drink, doogh, is popular in the Middle East between Lebanon, Iran and Afghanistan; it differs from ayran by the addition of herbs, usually mint, and is sometimes carbonated, commonly with carbonated water.

Lassi (urdu) is a yoghurt-based beverage originally from the Indian subcontinent that is usually slightly salty or sweet. Lassi is a staple of Punjab. In some parts of the subcontinent, the sweet version may be commercially flavoured with rosewater, mango or other fruit juice to create a very different drink. Salty lassi is usually flavoured with ground, roasted cumin and red chillies; this salty variation may also use buttermilk, and is interchangeably called *ghol* (Bengal), *mattha* (North India), *majjiga* (Andhra Pradesh), *moru* (Tamil Nadu and Kerala), *Dahi paani* (Odisha), *tak* (Maharashtra), or *chaas* (Gujarat). Lassi is also very widely drunk in Pakistan.

Sweetened yoghurt drinks are the usual form in Europe (including the UK) and the US, containing fruit and added sweeteners. These are typically called "drinking / drinkable yoghurt", such as Yop and BioBest Smoothie. Also available are "yoghurt smoothies", which contain a higher proportion of fruit and are more like smoothies. In Ecuador, yoghurt smoothies flavoured with native fruit are served with pan de yuca as a common type of fast food.

Lassi

Lassi is a popular and traditional Punjabi yogurt-based drink of India and Pakistan. It is made by blending yogurt with water or milk and Indian spices. Traditional lassi (also known as salted lassi, or, simply lassi) is a savory drink sometimes flavoured with ground roasted cumin while sweet lassi on the other hand is blended with sugar or fruits instead of spices. In Dharmic religions, yogurt sweetened with honey is used while performing religious rituals. Less common is lassi served with milk and is topped with a thin layer of clotted cream. Lassis are enjoyed chilled as a hot-weather refreshment, mostly taken with lunch. With a little turmeric powder mixed in, it is also used as a folk remedy for gastroenteritis.

Variations

Traditional Mild Salted Lassi

This form of lassi is more common in villages of Punjab & Porbandar, Gujarat (India). It is prepared by blending yogurt with water and adding salt and other spices to taste. The resulting beverage is known as salted lassi.

Sweet Lassi

Sweet lassi is a form of lassi flavoured with sugar, rosewater and/ or lemon, strawberry or other fruit juices. Saffron lassis, which are particularly rich, are a specialty of Sindh in Pakistan and Jodhpur and Rajasthan in India. *Makkhaniya lassi* is simply lassi with lumps of butter in it (*makkhan* is the Punjabi, Urdu, Hindi and Gujarati word for butter). It is usually creamy like a milkshake. Drinking sweet lassi can cause drowsiness, which might help people with sleep disorders.

Mango Lassi

Mango lassi is most commonly found in India and Pakistan though it is gaining popularity worldwide. It is made from yogurt, milk or water and mango pulp. It may be made with or without additional sugar. It is widely available in UK, Malaysia and Singapore, due to the sizable Pakistani/Indian minority, and in many other parts of the world. In various parts of Canada, mango lassi is a cold drink consisting of sweetened kesar mango pulp mixed with yogurt, cream, or ice cream. It is served in a tall glass with a straw, often with ground pistachio nuts sprinkled on top.

Bhang Lassi

Bhang (or bhung) lassi is a special lassi that contains *bhang*, a liquid derivative of cannabis (marijuana), which has effects similar to other eaten forms of marijuana. It is legal in many parts of India and mainly sold during Holi, when pakoras containing bhang are also sometimes eaten. Rajasthan is known to have licensed bhang shops, and in many places one can buy bhang products and drink bhang lassis. However, the term "bhang lassie" is more often a misnomer, as bhang is almost always drunk with thandai, which does not include any curd (joghurt), but either milk, water, crushed ice, sugar, kesar pista flavouring (a ready made thandai syrup), saffron, ground almonds, spices, gound melon and poppy seeds and bhang...etc.

Chaas

Chaas or chaach is a salted drink like lassi; however, chaas contains more water than lassi and has the butterfat removed, so its consistency is not as thick as lassi. Salt and Jeera (cumin seeds) are normally added for taste and sometimes even fresh coriander. Fresh ground ginger & green chillies may also be added as seasoning. Chaas is popular in the Pakistani/Indian Punjab regions of Bhakkar, Jalandhar, and D.I. Khan, and in the Indian states Gujarat and Rajasthan, where it is drunk with the main meal. It is known to aid digestion and is an excellent coolant in the Pakistani and Indian summers. It is called 'majjige' in Kannada, 'majjiga' in Telugu and 'moru' in Tamil and Malayalam.

Ayran

A drink in Turkey is similar to Lassi called Ayran. It is also made with yogurt and water. In Iran and Afghanistan a similar drink is called Doogh.

Tahn

Tahn is the Armenian version of the yogurt drink, consumed cold, with or without food. It is diluted with water and is flavoured with salt.

Cultural References

A 2008 print and television ad campaign for HSBC, written by Jeffree Benet of JWT Hong Kong, tells a tale of a Polish washing machine manufacturer's representative sent to India to discover why their sales are so high there. On arriving, the representative investigates a Lassi parlour, where he is warmly welcomed, and finds several washing machines being used to mix Lassi. The owner tells him he is able to "make ten times as much Lassi as I used to!"

On his *No Reservations* television program, celebrity chef Anthony Bourdain visited a "Govt Authorised" Bhang Shop in Jaisalmer Fort, Rajasthan. The proprietor offered him three varieties of bhang lassi: "normally strong, super duper sexy strong, and full power 24 hour, no toilet, no shower."

Buttermilk

Buttermilk refers to a number of dairy drinks. Originally, buttermilk was the liquid left behind after churning butter out of cream. It also refers to a range of fermented milk drinks, common in warm climates (e.g., Middle East, Pakistan, India, or the Southern United States) where fresh milk would otherwise sour quickly. It is also popular in Scandinavia and the Netherlands, despite the colder climates.

Buttermilk may also refer to a fermented dairy product produced from cow's milk with a characteristically sour taste caused by lactic acid bacteria. This variant is made in one of two ways:*cultured* buttermilk is made by adding lactic acid bacteria (*Streptococcus lactis*) to milk; *Bulgarian buttermilk* is created with a different strain of bacteria called *Lactobacillus bulgaricus*, which creates more tartness.

Whether traditional or cultured, the tartness of buttermilk is due to the presence of acid in the milk. The increased acidity is primarily due to lactic acid, a by-product naturally produced by lactic acid bacteria while fermenting lactose, the primary sugar found in milk.

As lactic acid is produced by the bacteria, the pH of the milk decreases and casein, the primary protein in milk, precipitates causing the curdling or clabbering of milk. This process makes buttermilk thicker than plain milk. While both traditional and cultured buttermilk contain lactic acid, traditional buttermilk tends to be thinner whereas cultured buttermilk is much thicker.

Traditional Buttermilk

Originally, buttermilk was the liquid left over from churning butter from cream. Traditionally, before cream could be skimmed from whole milk, it was left to sit for a period of time to allow the cream and milk to separate. During this time, the milk would be fermented by the naturally occurring lactic acid-producing bacteria in the milk. This facilitates the butter churning process since fat from cream with a lower pH will coalesce more readily than that from fresh cream. The acidic environment also helps prevent potentially harmful microorganisms from growing, increasing shelf-life. However, in establishments that used cream separators, the cream would hardly be acid at all. In the Indian subcontinent, buttermilk is taken to be the liquid leftover after extracting butter from churned yogurt (dahi). Today, this is called *traditional buttermilk.* Traditional buttermilk is still common in many Indo-Pakistani households but rarely found in western countries. In Southern India and most areas of the Punjab, buttermilk with added water, sugar and/or salt, asafoetida, and curry leaves is given at stalls in festival times.

Cultured Buttermilk

Commercially available cultured buttermilk is pasteurized and homogenized (if 1% or 2% fat) milk which has been inoculated with a culture of lactic acid bacteria to simulate the naturally occurring bacteria found in the old-fashioned product. Some dairies add colored flecks of butter to cultured buttermilk to simulate the residual pieces of butter that can be left over from the churning process of traditional buttermilk. Buttermilk solids have increased in importance in the food industry. Such solids are used in ice cream manufacture. Adding specific strains of bacteria to pasteurized milk allows more consistent production.

In the early 1900s, cultured buttermilk was labelled *artificial buttermilk,* to differentiate it from traditional buttermilk, which was known as *natural* or *ordinary buttermilk.*

Acidified buttermilk is a related product that is made by adding a food-grade acid (such as lemon juice) to milk.

Probiotics

Probiotics are live microorganisms thought to be beneficial to the host organism. According to the currently adopted definition by FAO/ WHO, probiotics are: "Live microorganisms which when administered in adequate amounts confer a health benefit on the host". Lactic acid bacteria (LAB) and bifidobacteria are the most common types of microbes used as probiotics; but certain yeasts and bacilli may also be helpful. Probiotics are commonly consumed as part of fermented foods with specially added active live cultures; such as in yogurt, soy yogurt, or as dietary supplements.

Etymologically, the term appears to be a composite of the Latin preposition *pro* ("for") and the Greek adjective biotic, the latter deriving from the noun *âβïò* (bios, "life").

At the start of the 20th century, probiotics were thought to beneficially affect the host by improving its intestinal microbial balance, thus inhibiting pathogens and toxin producing bacteria. Today, specific health effects are being investigated and documented including alleviation of chronic intestinal inflammatory diseases, prevention and treatment of pathogen-induced diarrhea, urogenital infections, and atopic diseases.

History

The original observation of the positive role played by certain bacteria was first introduced by Russian scientist and Nobel laureate Eli Metchnikoff, who in the beginning of the 20th century suggested that it would be possible to modify the gut flora and to replace harmful microbes with useful microbes. Metchnikoff, at that time a professor at the Pasteur Institute in Paris, produced the notion that the aging process results from the activity of putrefactive (proteolytic) microbes producing toxic substances in the large bowel.

Proteolytic bacteria such as clostridia, which are part of the normal gut flora, produce toxic substances including phenols, indols and ammonia from the digestion of proteins. According to Metchnikoff these compounds were responsible for what he called "intestinal auto-intoxication", which caused the physical changes associated with old age.

It was at that time known that milk fermented with lactic-acid bacteria inhibits the growth of proteolytic bacteria because of the low pH produced by the fermentation of lactose. Metchnikoff had also observed that certain rural populations in Europe, for example in Bulgaria and the Russian steppes who lived largely on milk fermented by lactic-acid bacteria were exceptionally long lived.

Based on these facts, Metchnikoff proposed that consumption of fermented milk would "seed" the intestine with harmless lactic-acid bacteria and decrease the intestinal pH and that this would suppress the growth of proteolytic bacteria. Metchnikoff himself introduced in his diet sour milk fermented with the bacteria he called "Bulgarian Bacillus" and found his health benefited. Friends in Paris soon followed his example and physicians began prescribing the sour milk diet for their patients.

Bifidobacteria were first isolated from a breast-fed infant by Henry Tissier who also worked at the Pasteur Institute. The isolated bacterium named Bacillus bifidus communis was later renamed to the genus *Bifidobacterium*. Tissier found that bifidobacteria are dominant in the gut flora of breast-fed babies and he observed clinical benefits from treating diarrhea in infants with bifidobacteria. The claimed effect was bifidobacterial displacement of proteolytic bacteria causing the disease.

During an outbreak of shigellosis in 1917, German professor Alfred Nissle isolated a strain of *Escherichia coli* from the feces of a soldier who was not affected by the disease. Methods of treating infectious diseases were needed at that time when antibiotics were not yet available, and Nissle used the *Escherichia coli* Nissle 1917 strain in acute gastrointestinal infectious salmonellosis and shigellosis.

In 1920, Rettger demonstrated that Metchnikoff's "Bulgarian Bacillus", later called *Lactobacillus delbrueckii subsp. bulgaricus*, could not live in the human intestine, and the fermented food phenomena petered out. Metchnikoff's theory was disputable (at this stage), and people doubted his theory of longevity.

After Metchnikoff's death in 1916, the centre of activity moved to the United States. It was reasoned that bacteria originating from the gut were more likely to produce the desired effect in the gut, and in 1935 certain strains of *Lactobacillus acidophilus* were found to be very active when implanted in the human digestive tract. Trials were carried out using this organism, and encouraging results were obtained especially in the relief of chronic constipation.

The term "probiotics" was first introduced in 1953 by Kollath. Contrasting antibiotics, probiotics were defined as microbially derived factors that stimulate the growth of other microorganisms. In 1989 Roy Fuller suggested a definition of probiotics which has been widely used: "*A live microbial feed supplement which beneficially affects the host animal by improving its intestinal microbial balance*". Fuller's definition emphasizes the requirement of viability for probiotics and introduces the aspect of a beneficial effect on the host.

In the following decades intestinal lactic acid bacterial species with alleged health beneficial properties have been introduced as probiotics, including *Lactobacillus rhamnosus, Lactobacillus casei,* and *Lactobacillus johnsonii.*

Benefits

Experiments into the benefits of probiotic therapies suggest a range of potentially beneficial medicinal uses for probiotics. For many of the potential benefits, research is limited and only preliminary results are available. It should be noted that the effects described are *not* general effects of probiotics. Recent research on the molecular biology and genomics of *Lactobacillus* has focused on the interaction with the immune system, anti-cancer potential, and potential as a biotherapeutic agent in cases of antibiotic-associated diarrhoea, travellers' diarrhoea, pediatric diarrhoea, inflammatory bowel disease and irritable bowel syndrome. All effects can only be attributed to the individual strain(s) tested. Testing of a supplement does not indicate benefit from any other strain of the same species, and testing does not indicate benefit from the whole group of LAB (or other probiotics).

Managing Lactose Intolerance

As lactic acid bacteria actively convert lactose into lactic acid, ingestion of certain active strains may help lactose intolerant individuals tolerate more lactose than they would have otherwise.

Preventing Colon Cancer

In laboratory investigations, some strains of LAB (*Lactobacillus bulgaricus*) have demonstrated anti-mutagenic effects thought to be due to their ability to bind with heterocyclic amines, which are carcinogenic substances formed in cooked meat. Animal studies have demonstrated that some LAB can protect against colon cancer in rodents, though human data is limited and conflicting. Most human trials have found that the strains tested may exert anti-carcinogenic effects by decreasing the activity of an enzyme called α-glucuronidase (which can generate carcinogens in the digestive system). Lower rates of colon cancer among higher consumers of fermented dairy products have been observed in one population study.

Lowering Cholesterol

Animal studies have demonstrated the efficacy of a range of LAB to be able to lower serum cholesterol levels, presumably by breaking down bile in the gut, thus inhibiting its reabsorption (which enters the blood as cholesterol). Some, but not all human trials have shown that dairy foods fermented with specific LAB can produce modest reductions in total and LDL cholesterol levels in those with normal levels to begin with, however trials in hyperlipidemic subjects are needed.

Lowering Blood Pressure

Several small clinical trials have indicated that consumption of milk fermented with various strains of LAB may result in modest reductions in blood pressure. It is thought that this is due to the ACE inhibitor-like peptides produced during fermentation.

Improving Immune Function and Preventing Infections

LAB are thought to have several presumably beneficial effects on immune function. They may protect against pathogens by means of competitive inhibition (i.e., by competing for growth) and there is evidence to suggest that they may improve immune function by increasing the number of IgA-producing plasma cells, increasing or improving phagocytosis as well as increasing the proportion of T lymphocytes and Natural Killer cells.

Clinical trials have demonstrated that probiotics may decrease the incidence of respiratory tract infections and dental caries in children. LAB foods and supplements have been shown to aid in the treatment and prevention of acute diarrhea, and in decreasing the severity and duration of rotavirus infections in children and travelers' diarrhea in adults.

A 2010 study suggested that the anecdotal benefits of probiotic therapies as beneficial for preventing secondary infections, a common complication of antibiotic therapy, may be because keeping the immune system primed by eating foods enhanced with "good" bacteria may help counteract the negative effects of sickness and antibiotics. It was thought that antibiotics may turn the immune system "off" while probiotics turns it back on "idle", and more able to quickly react to new infections.

Helicobacter Pylori

LAB are also thought to aid in the treatment of *Helicobacter pylori* infections (which cause peptic ulcers) in adults when used in combination with standard medical treatments. However more studies are required into this area.

Antibiotic-associated Diarrhea

Antibiotic-associated diarrhea (AAD) results from an imbalance in the colonic microbiota caused by antibiotic therapy. Microbiota alteration changes carbohydrate metabolism with decreased short-chain fatty acid absorption and an osmotic diarrhea as a result. Another consequence of antibiotic therapy leading to diarrhea is overgrowth of potentially pathogenic organisms such as *Clostridium difficile*.

Probiotic treatment can reduce the incidence and severity of AAD as indicated in several meta-analyses. However, further documentation of these findings through randomized, double blind, placebo-controlled trials are warranted.

Efficacy of probiotic AAD prevention is dependent on the probiotic strain(s) used and on the dosage. Up to a 50% reduction of AAD occurrence has been found. No side-effects have been reported in any of these studies. Caution should, however, be exercised when administering probiotic supplements to immunocompromised individuals or patients who have a compromised intestinal barrier.

Reducing Inflammation

LAB and supplements have been found to modulate inflammatory and hypersensitivity responses, an observation thought to be at least in part due to the regulation of cytokine function. Clinical studies suggest that they can prevent reoccurrences of inflammatory bowel disease in adults, as well as improve milk allergies. They are not effective for treating eczema, a persistent skin inflammation. How probiotics counteract immune system overactivity remains unclear, but a potential mechanism is desensitization of T lymphocytes, an important component of the immune system, towards pro-inflammatory stimuli .

Improving Mineral Absorption

It is hypothesized that probiotic lactobacilli may help correct malabsorption of trace minerals, found particularly in those with diets high in phytate content from whole grains, nuts, and legumes.

Preventing Harmful Bacterial growth Under Stress

In a study done to see the effects of stress on intestinal flora, rats that were fed probiotics had little occurrence of harmful bacteria latched onto their intestines compared to rats that were fed sterile water.

Treating Irritable Bowel Syndrome and Colitis

B. infantis 35624, sold as Align, was found to improve some symptoms of irritable bowel syndrome in women in a recent study. Another probiotic bacterium, *Lactobacillus plantarum* 299v, was also found to be effective in reducing IBS symptoms. Additionally, a probiotic formulation, VSL#3, was found to be safe in treating ulcerative colitis, though efficacy in the study was uncertain.

Bifidobacterium animalis DN-173 010 may help. For maintenance of remission of ulcerative colitis, Mutaflor (*E.coli* Nissle 1917) there are 3 controlled, randomized, double blind clinical studies which have proven equivalence of Mutaflor and mesalazine (5-ASAs).

Managing Urogenital Health

Several in vitro studies have revealed probiotics' potential in relieving urinary tract infections and bacterial vaginosis. Results have been varied on these studies, and in vivo studies are still required in this area to determine efficacy.

Other

A study in 2004 testing the immune system of students given either milk or Actimel over a 6 week exam period (3 weeks of studying, 3 weeks of exams) tested 19 different biomarkers. Of these 19 biomarkers only 2 were shown to be different between the two groups, increased production of lymphocytes and increased production of CD56 cells. The tests were not blind and show that certain probiotic strains may have no overall effect on the immune system or on its ability.

A 2007 study at University College Cork in Ireland showed that a diet including milk fermented with *Lactobacillus* bacteria prevented *Salmonella* infection in pigs. A 2007 clinical study at Imperial College London showed that preventive consumption of a commercially available probiotic drink containing *L casei DN-114001*, *L bulgaricus*, and *S thermophilus* can reduce the incidence of antibiotic-associated diarrhea and *C difficile*-associated diarrhea.

The efficacy and safety of a daily dose of *Lactobacillus acidophilus* CL1285 in the prevention of AAD was demonstrated by Montreal's Maisonneuve-Rosemont Hospital, in a clinical study of hospitalized patients. Current research is focusing on the molecular biology and genomics of *Lactobacillus* and bifidobacteria. The application of modern whole genome approaches is providing insights into bifidobacterial evolution, while also revealing genetic functions that explain their presence in the particular ecological environment of the gastrointestinal tract. Probiotics are used in industry to improve yields of pork and chicken production.

Disadvantages

In some situations, such as where the person consuming probiotics is critically ill, probiotics could be harmful. In a therapeutic clinical trial conducted by the Dutch Pancreatitis Study Group, the consumption of a mixture of six probiotic bacteria, increased the death rate of patients with predicted severe acute pancreatitis.

In a clinical trial conducted at the University of Western Australia, aimed at showing the effectiveness of probiotics in reducing childhood allergies, Dr Susan Prescott and her colleagues gave 178 children either a probiotic or a placebo for the first six months of their life. Those given the good bacteria were more likely to develop a sensitivity to allergens. Some hospitals have reported treating lactobacillus septicaemia, which is a potentially fatal disease caused by the consumption of probiotics by people with lowered immune systems or who are already very ill.

There is no published evidence that probiotic supplements are able to replace the body's natural flora when these have been killed off; indeed bacterial levels in feces disappear within days when supplementation ceases. Probiotics taken orally can be destroyed by the acidic conditions of the stomach. A number of micro-encapsulation techniques are being developed to address this problem.

Recent studies indicate that probiotic products such as yogurts could be a cause for obesity trends. However, this is contested as the link to obesity and other health related issue with yogurt may link to its dairy attributes. Some experts are skeptical on the efficacy of many strains and believe not all subjects will benefit from the use of probiotics. A criticism of probiotic supplements is the cost and value of probiotics products.

Bibliography

Alec, William: *The Dairy Chemical Industry*, London: Longman Group Limited, 1971.

Anifantakis, EM.: *Greek Cheeses*, Nat. Dairy Committee of Greece Publ., Athens, Greece, 1991.

Arora, Dinesh: *Biotech's Dictionary of Dairy Science*, Biotech Books, Delhi, 296.

Ashby, E.: *Technology and the Academics*, London: Macmillan, 1959.

Barbeau, G.: *Tropical Fruits in Nicaragua*, Managua, Nicaragua Ministerio de Desarrollo Agropecuario, Agraria, 1990.

Basavaraj, S. Benni: *Dairy Co-operative Management and Practice*, Rawat, Delhi, 2005.

Bhattacharya, Lata: *Biochemistry of Nutrition*, Discovery, Delhi, 2010.

Bohra, Babita: *Dairy Farming in Mountain Areas*, Daya, Delhi, 2006.

Brock, H.: *History of Dairy Chemistry*, New York: Norton, 1992.

Bucciarelli, L.: *Designing Engineers in Dairy* , Cambridge: MIT Press, 1995.

Bushnell, R.B.: *Dry Cow Feeding and Management*, A Western Regional Extension Publication, 1979.

Chand, Ram: *Decision-Making of Dairy Beneficiaries: Role of Aspiration, Motivation and Knowledge*, Om Pub, Delhi, 2010.

Chhazllani, V K: *Dairy Chemistry and Animal Nutrition*, Manglam Pub, Delhi, 2008.

Cohen, Lizabeth, *Making a New Deal, Industrial Workers in Chicago, 1919-1939* Cambridge University Press, 1991.

Damasio, AR.: *Descarte's Error: Emotion, Reason and the Human Brain*, New York, 1994.

David, M.: *Ideas in Chemistry: A History of the Science*, New Brunswick, N.J.: Rutgers University Press, 1992.

De, Sukumar: *Outlines of Dairy Technology*, Oxford University Press, Delhi, 2001.

Deepti, R. Haritash and Tej Verma: *Promoting Community Nutrition Through Participatory Monitoring and Evaluation*, Advance, 2001.

Desmond, Glenn, and John A, Marcell, *Handbook of Small Business Valuation Formulas and Rules of Thumb,* Valuation Press, 1993.

Devraj, B: *Impact of Dairy On Small and Marginal Farmers*, Prateeksha Publications, Delhi, 2010.

Droop, H. Richmond: *Laboratory Manual of Dairy Analysis,* Biotech, 2004.

Fox, Patrick F.: *Advanced Dairy Chemistry: Proteins*, New York: Elsevier Applied Science, 1992.

Friberg, Stig. E..: *Food Emulsions*, New York: M. Dekker, 1997.

Gautam, U.S. and Ram Chand: *Decision-Making of Dairy Beneficiaries: Role of Aspiration, Motivation and Knowledge*, Om Pub, Delhi, 2010.

Gowdy, J., *Coevolution Economics: The Economy, Society and the Environment*, Kluwer, Boston, 1994.

Guarti, Luigi, *The Valuation of Firms*, Blackwell Publishing, 1994.

Gupta, Sudhir: *Hand Book of Dairy Formulations, Processes and Milk Processing Industries*, Engineers India Res Inst, Delhi, 2003.

Hemantaranjan, A.: *Advancements in Iron Nutrition Research*, Scientific, Delhi, 1995.

Hornig, Susanna: *A Grain of Truth: The Media, the Public, and Biotechnology*, Lantham, MD: Rowman and Littlefield, 2001.

Hui, Yiu H.: *Dairy Science and Technology Handbook*, New York: Wiley, 1993.

Jensen, Robert G.: *Handbook of Milk Composition*, San Diego: Academic Press, 1995.

Jensen, Robert G.: *Handbook of Milk Composition*, San Diego: Academic Press, 1995.

Jha, S. N.: *Dairy and Food Processing Plant Maintenance: Theory and Practice,* International Book Distributing, Delhi, 2006.

John A, Marcell, *Handbook of Small Business Valuation Formulas and Rules of Thumb,* Valuation Press, 1993.

Julia, F.: *Fruits of Warm Climates*, Miami, Julia F. Morton Publisher, 1987.

Kango, Mangala: *Normal Nutrition: Fundamental and Management*, RBSA, Delhi, 2003.

Kapoor, Ajay: *Dairy Science and Technology,* Vishvabharti Pub, Delhi, 2005.

Khemka, Devyani: *Animal Physiology*, Dominant, Delhi, 2003.

King, C.J.: *Separation Processes*, McGraw Hill, 1980.

Koli, P. A.: *Dairy Development in India: Challenges Before Co-Operatives*, Shruti Pub, Delhi, 2007.

Kosikowski, Frank, and Vikram V. Mistry: *Cheese and FermentedMilk Foods*, Great Falls, Va.: Kosikowski, 1997.

Lata Bhattacharya: *Biochemistry of Nutrition*, Discovery, Delhi, 2010.

Law, Barry A.: *Microbiology and Biochemistry of Cheese and Fermented Milk*, London: Blackie Academic & Professional, 1997.

March, J. G.: *A Behavioural Theory of the Firm*, Englewood-Cliffs, Prentice Hall, 1963.

Marth, Elmer H., and James L. Steele: *Applied Dairy Microbiology*, New York: M. Dekker, 1998.

Mayrose, V.B.: *Handbook of livestock Management Techniques*, Macmillan Publishing Co., New York, 1981.

Meyer, M. W.: *Theory of Organizational Structure*, Indianapolis, 1977.

Michael, J. Lewis,: *SeparationProcesses in the Food and Biotechnology Industries*, Cambridge, U.K.: Woodhead, 1996.

Miller, Gregory D., Judith K. Jarvis, and Lois D. McBean.: *Handbook of Dairy Foods and Nutrition*, Boca Raton, Fla.: CRC Press, 1999.

Mowery, D. C.: *Technology and Wealth of Nations*, Stanford: Stanford University Press, 1992.

Mudgal, V. D.; K. K. Singhal and D. D. Sharma: *Advances in Dairy Animal Production,* International Book Distribut, 2003.

Myers, R.: *Basics of Dairy Chemistry*, Atlantic, Delhi, 2007.

Nedderman, R. M.: *A Handbook of Unit Operations*, Academic, London, 1971.

Nisha, Maimun: *Health, Food and Nutrition*, Kalpaz, Delhi, 2006.

Pal, Kanwar Singh Sangwan: *Technology of Dairy Plant Operations*, Agrobios, Delhi, 2008.

Pandey, D. N. and Amita Bajpai: *Recent Trends in Animal Nutrition and Feed Technology for Livestock, Pets and Laboratory Animals*, International, 2003.

Parihar, Pradeep and Leena Parihar: *Dairy Microbiology*, Agrobios, Delhi, 2008.

Patton, Stuart: *Principles of Dairy Chemistry*, Huntington, N.Y.: Krieger, 1976.

Pirtle, Thomas Ross: *History of the Dairy Industry,* Chicago: Mojonnier Bros. Company, 1926.

Porter, A.R., J.A. Sims, and C.F. Foreman: *Dairy Cattle in American Agriculture,* Iowa, Iowa State University Press, 1965.

Prince, John: *Dairy Farming: Being the Theory, Practice, and Methods of Dairying*, New York, 1888.

Qystein, V. Sjaastad: *Physiology of Domestic Animals*, International Book Distributing Co., Delhi, 2005.

Rao, M K: *Food and Dairy Microbiology*, Manglam Pub, Delhi, 2007.

Rao, P. Venkateshwara: *Dairy Farm Business Management*, Biotech Books, Delhi, 2008.

Samvel, A. P. V.: *Agri-Business Management*, Satish Serial Pub, Delhi, 2008.

Sarkar, A: *Advanced Organic Chemistry: Reactions and Mechanisms*, Swastik Publications, Delhi, 2011.

Sharma, Ramakant: *Chemical and Microbiological Analysis of Milk and Milk Products,* International Book Distributing, 2006.

Sheldon, John Prince: *Dairy Farming: Being the Theory, Practice, and Methods of Dairying*, New York, 1888.

Shukla , Arvind N.: *Textbook of Dairy Chemistry,* Discovery Pub, Delhi 2010.

Shukla, N.: *Textbook of Dairy Chemistry,* Discovery Pub, Delhi 2010.

Singh, G.: *Chemistry of Amino-Acids and Proteins*, Discovery, Delhi, 2007.

Singh, Harmeet: *Dairy Farming*, APH, Delhi, 2005.

Singh, Vir and Babita Bohra: *Dairy Farming in Mountain Areas*, Daya, Delhi, 2006.

Singhal, K. K. and D. D. Sharma: *Advances in Dairy Animal Production,* International Book Distribut, 2003.

Spreer, Edgar: *Milk and Dairy Product Technology*, Translated by Axel Mixa. New York: M. Dekker, 1998.

Sukumar De: *Outlines of Dairy Technology*, Oxford University Press, Delhi, 2001.

Suresh, M: *Dairy Development and Income Distribution in India*, Abhijeet Pub, Delhi, 2007.

Taylor, F. W.: *The Principles of Scientific Management*, New York, 1917.

Thompson, Paul B.: *Food Biotechnology in Ethical Perspective*, Aspen, CO: Aspen Publishers, 1997.

Thomson, Sutherland: *Grading Dairy Produce*, Medi World Press, Delhi, 1995.

Tomar, S.K. and Gunjan Goel: *Applied Dairy and Food Microbiology,* Agrotech, 2005.

Tyagi, Prasum: *A Textbook of Animal Physiology*, Dominant, Delhi, 2010.

Upadhyay, K.G. and Vyas, S.H.: *Composition of Camel's Milk*, Gujarat Agric. University, 1982.

van, M.A. J. S. Boekel: *Dairy Technology: Principles of Milk Properties and Processes,* New York: M. Dekker, 1999.

Walstra, Pieter, and Robert Jenness: *Dairy Chemistry and Physics*, New York: Wiley, 1984.

Weimar, Mark R. and Don P. Blayney: *Landmarks in the U.S. Dairy Industry*, USDA, ERS, Agriculture Information Bulletin Number 694, 1994.

Yegge, Wilbur M., *A Basic Guide for Valuing a Company*, New York, Wiley, 1996.

Index

K

L

M

N

O

P

R

S

T

V

W

Y

❑❑❑